Directions in Partial Differential Equations

Publication No. 54
of the Mathematics Research Center
The University of Wisconsin–Madison

Directions in Partial Differential Equations

Edited by

Michael G. Crandall, Paul H. Rabinowitz, Robert E. L. Turner

Mathematics Research Center
The University of Wisconsin–Madison
Madison, Wisconsin

Proceedings of a Symposium
Conducted by the Mathematics Research Center
The University of Wisconsin–Madison
October 28–30, 1985

ACADEMIC PRESS, INC.
Harcourt Brace Jovanovich, Publishers

Boston Orlando San Diego
New York Austin London Sydney
Tokyo Toronto

ACADEMIC PRESS, INC.
Orlando, Florida 32887

United Kingdom Edition published by
ACADEMIC PRESS INC. (LONDON) LTD.
24-28 Oval Road, London NW1 7DX

Library of Congress Cataloging-in-Publication Data

Directions in partial differential equations.

(Publication no. 54 of the Mathematics Research Center, the University of Wisconsin—Madison)
"Proceedings of a symposium conducted by the Mathematics Research Center, the University of Wisconsin—Madison, October 28–30, 1985."
Includes index.
1. Differential equations, Partial—Congresses.
I. Crandall, Michael G. II. Rabinowitz, Paul H.
III. Turner, Robert E. L. IV. University of Wisconsin—Madison. Mathematics Research Center. V. Series: Publication . . . of the Mathematics Research Center, the University of Wisconsin—Madison; no. 54.
QA3.U45 no. 54 510 s [515.3'53] 86-32167
[QA377]
ISBN 0-12-195255-X (alk. paper)

87 88 89 90 9 8 7 6 5 4 3 2 1
Printed in the United States of America

This volume is dedicated to Professor John A. Nohel in recognition of his 60th birthday and in appreciation of his many contributions to mathematics both through his leading scientific work and through his warm and energetic service to the mathematical community as director of the Mathematics Research Center.

Contents

Preface

This volume collects written versions of the lectures given at a symposium on partial differential equations conducted by the Mathematics Research Center of the University of Wisconsin–Madison in October 1985. One of the goals of the conference was to exhibit the remarkable vitality and breadth of current activity in partial differential equations as well as the high scientific level of ongoing work in the field. The distinguished slate of speakers rose to this task admirably and all but two were able to contribute manuscripts. In this volume, the reader will find fascinating contributions to the calculus of variations, differential geometry, the study of development of singularities, regularity theory, hydrodynamics, mathematical physics, asymptotic behaviour, critical point theory, the use of probabilistic methods, the modern theory of Hamilton–Jacobi equations, the interaction between theory and numerical methods for partial differential equations, and recent attempts to understand oscillatory phenomena in solutions of nonlinear equations. The reader who explores these papers will learn new tools and ideas, new results, interesting points of view, and important problems awaiting resolution.

The organizing committee, which consists of the editors and Andrew Majda, wishes to express its appreciation to the United States Army for its support of the symposium under Contract No. DAAG29-80-C-0041 and the National Science Foundation under Grant No. DMS-82-10950, #4. We also wish to thank Gladys Moran for her usual expert handling of the symposium and Sally Ross for her crucial assistance in producing this volume.

Symposium Speakers

J. M. Ball
Haïm Brezis
Luis A. Caffarelli
Phillip Colella
Ronald J. DiPerna
Ciprian Foias
Avner Friedman
James M. Hyman
S. Klainerman
P.-L. Lions
G. Papanicolaou
Daniel W. Stroock
Luc C. Tartar
Clifford Taubes
Shing-Tung Yau

Contributors

Numbers in parentheses indicate pages on which authors' contributions begin.

J.M. Ball (1), *Department of Mathematics, Heriot–Watt University, Riccarton, Edinburgh EH14 4AS, Scotland, U.K.*

Haïm Brezis (17), *Département de Mathématiques, Université P. et M. Curie, 4, pl. Jussieu, 75252 Paris Cédex 05, France*

Luis A. Caffarelli (37), *Department of Mathematics, University of Chicago, Chicago, Illinois 60637*

Ronald J. DiPerna (43), *Department of Mathematics, University of California, Berkeley, California 94720*

Ciprian Foias (55), *Mathematics Department, Indiana University, Bloomington, Indiana 47405*

Avner Friedman (75), *Department of Mathematics, Purdue University, West Lafayette, Indiana 47907*

James M. Hyman (89), *Center for Nonlinear Studies, Theoretical Division, MS B284, Los Alamos National Laboratory, Los Alamos, New Mexico 87545*

S. Klainerman (113), *Courant Institute of Mathematical Sciences, New York University, 251 Mercer Street, New York, New York 10012*

B. LeMesurier (159), *Department of Mathematical Sciences, Rensselaer Polytechnic Institute, Troy, New York 12180*

P.-L. Lions (145), *Ceremade, Université Paris–Dauphine, Place de Lattre de Tassigny, 75775 Paris Cédex 16, France*

Basil Nicolaenko (89), *Center for Nonlinear Studies, Theoretical Division, MS B284, Los Alamos National Laboratory, Los Alamos, New Mexico 87545*

G. Papanicolaou (159), *Courant Institute of Mathematical Sciences, New York University, 251 Mercer Street, New York, New York 10012*

Richard Schoen (235), *Department of Mathematics, University of California, San Diego, La Jolla, California 92093*

Daniel W. Stroock (203), *Department of Mathematics, Massachusetts Institute of Technology, Cambridge, Massachusetts 02139*

C. Sulem (159), *Department of Mathematics and Computer Science, Ben Gurion University of the Negev, P. O. Box 653 Beer-Sheva, 84 105, Israel*

P.-L. Sulem (159), *Department of Mathematics, Tel Aviv University, Ramat Aviv, Israel*

Luc C. Tartar (211), *Centre d'Etudes de Limeil–Valenton, DMA/MCN, BP. 27, 94190 Villeneuve Saint Georges, France*

Roger Témam (55), *Université de Paris–Sud, Mathématiques, Bât. 425, 91405 Orsay Cédex, France*

Shing-Tung Yau (235), *Department of Mathematics, University of California–San Diego, La Jolla, California 92093*

Singular Minimizers and their Significance in Elasticity

J. M. Ball

1. INTRODUCTION.

Minimizers of integrals of the calculus of variations typically possess singularities. For problems arising from mechanics, such singularities may represent physically interesting instabilities. We explore this here in the context of elasticity theory, for which a complete classification of possible singularities is not known. Certainly, as will be described below, singular minimizers exist that model aspects of solid phase transformations and certain modes of fracture. But it remains to be seen if certain singularities encountered elsewhere in the calculus of variations can occur in elasticity, or whether these are eliminated as a consequence of low spatial dimensions and invariance requirements. Perhaps some such singularities are already in the experimental literature for those with the eyes to see them.

The plan of the paper is as follows. In §2 the basic problem of energy minimization in elasticity is described, together with a bare minimum of information concerning properties of the stored-energy function. In each of the subsequent sections a particular type of singularity is discussed, the order being roughly that of increasing degree of singularity.

ISBN 0-12-195255-X

2. MINIMIZATION OF THE ENERGY IN ELASTICITY.

Consider an elastic body, which in a reference configuration occupies the bounded domain Ω of $\mathbb{R}^n$. We suppose that the boundary $\partial\Omega$ of Ω is sufficiently smooth. Points in Ω are denoted by $x = (x^1,\dots,x^n)$. A deformation of the body is a mapping $u : \Omega \to \mathbb{R}^n$. The corresponding total elastic energy is given by

$$I(u) = \int_\Omega W(x, Du(x))\,dx, \tag{2.1}$$

where W is the stored-energy function and Du(x) denotes the gradient of u at x. The deformation u is required to satisfy appropriate boundary conditions; as a typical example we consider the boundary conditions

$$u\Big|_{\partial\Omega_1} = f, \tag{2.2}$$

where $\partial\Omega_1$ is a given subset (possibly empty) of $\partial\Omega$ and $f : \partial\Omega_1 \to \mathbb{R}^n$ is a given function.

Thermodynamical reasoning (cf Duhem [20], Ericksen [21], Ball [7]) suggests that equilibrium configurations are minimizers or limits of minimizing sequences for I subject to (2.2). Here we have for simplicity neglected external body and surface forces. More significantly, perhaps, we have ignored contributions to the energy due to interior point, line or surface discontinuities; these can be of importance in controlling some of the singularities discussed here, although in many cases the effects are small. The explicit dependence of W on x may arise from inhomogeneity of the body (i.e. the material is different at different points), from non-uniform composition of a body composed of a single material, or from the presence of a temperature gradient (Ball & Knowles [10]).

We assume that the stored-energy function $W : \Omega \times M_+^{n\times n} \to \mathbb{R}$ is sufficiently regular. Here $M_+^{n\times n}$ denotes the subset of the space $M^{n\times n}$ of real $n \times n$ matrices consisting of those matrices with strictly positive determinant. With a view to ensuring that deformations u with finite energy are invertible and orientation-preserving it is frequently assumed that $W(x,A) \to \infty$ as $\det A \to 0+$. (For information concerning invertibility see Ball [4], Ciarlet & Nečas [16].) The dependence of W(x,A) on A is restricted by the frame-

<u>indifference</u> condition

$$W(x,QA) = W(x,A) \qquad \text{for all} \quad Q \in SO(n), \tag{2.3}$$

which asserts that the elastic energy of a body is unaffected by a rigid rotation. In addition W may satisfy material symmetries, such as the <u>isotropy</u> condition

$$W(x,AQ) = W(x,A) \qquad \text{for all} \quad Q \in SO(n). \tag{2.4}$$

It is well known (see [6] for a discussion and the classical references) that (2.3),(2.4) together imply that

$$W(x,A) = \Phi(x; v_1,\ldots,v_n), \tag{2.5}$$

for some function Φ that is invariant to permutations of the eigenvalues $v_i = v_i(A)$, $1 \leqslant i \leqslant n$, of $\sqrt{A^TA}$. These eigenvalues are usually called the <u>principal stretches</u>. Elastic crystals are not in general isotropic, but satisfy more complicated symmetry conditions related to their lattice structure.

In order to set up the minimization problem more precisely it is necessary to introduce a function space of admissible deformations. The choice of this function space involves in particular a choice for the meaning to be attached to Du, for a nonsmooth deformation u, which can dramatically affect the predictions of the model. The rationale for preferring one function space to another warrants further study. In this paper we consider the problem of minimizing I in the set

$$\mathcal{A}_p = \{u \in W^{1,p}\colon u\big|_{\partial\Omega_1} = f\}.$$

Here and below, for $1 \leqslant p \leqslant \infty$, $W^{1,p} = W^{1,p}(\Omega;\mathbb{R}^n)$ denotes the usual Sobolev space of mappings $u : \Omega \to \mathbb{R}^n$.

Note that minimizers u of I in $\mathcal{A}_p$ formally satisfy on $\partial\Omega \setminus \partial\Omega_1$ the natural boundary condition

$$\frac{\partial W}{\partial A}(x,DU)\; N(x) = 0, \tag{2.6}$$

where $N(x)$ denotes the unit outward normal to $\partial\Omega$ at x. This expresses the fact that the surface traction on $\partial\Omega$ at x is zero.

3. SMOOTH MINIMZERS.

For simplicity we consider in this section the case of a homogeneous material, so that $W = W(A)$ and

$$I(u) = \int_\Omega W(Du(x))dx. \tag{3.1}$$

(a) affine minimizers

An affine deformation $\bar{u}$ is one of the form

$$\bar{u}(x) = A_o x + a; \quad A_o \in M_+^{n\times n}, \quad a \in \mathbb{R}^n. \tag{3.2}$$

In the zero traction problem ($\partial\Omega_1$ empty) $\bar{u}$ given by (3.2) is clearly an absolute minimizer of I in $\mathcal{A}_p$ if and only if

$$W(A) \geqslant W(A_o) \qquad \text{for all} \quad A \in M_+^{n\times n}, \tag{3.3}$$

that is, A_o is an absolute minimizer of W.

In the pure displacement problem ($\partial\Omega_1 = \partial\Omega$) $\bar{u}$ given by (3.2) is by definition an absolute minimizer of I in $\mathcal{A}_p$ if and only if W is $W^{1,p}$-quasiconvex at A_o (see Ball & Murat [14]). Sufficient conditions for W to be $W^{1,1}$-quasiconvex (and hence $W^{1,p}$-quasiconvex whenever $1 \leqslant p \leqslant \infty$) at every $A_o \in M_+^{n\times n}$ are that

(i) W be polyconvex i.e. W can be written as a convex function of the minors of A (if $n = 3$, for example, the condition is that $W(A) = g(A, \operatorname{adj} A, \det A)$ for all $A \in M_+^{n\times n}$ and some convex function $g : M^{3\times 3} \times M^{3\times 3} \times (0,\infty) \rightarrow \mathbb{R}$), and

(ii) $W(A) \geqslant c_0 + c_1|A|^n$ for all $A \in M_+^{n\times n}$, where $c_1 > 0$ and c_0 are constants.

For the original definition and results concerning quasiconvexity see Morrey [38]. Existence theorems under polyconvexity conditions applying, for example, to realistic models of natural rubbers were given in [2]. There is a rapidly growing literature concerning quasiconvexity, and a number of relevant papers are cited in [8].

(b) smooth non-affine minimizers

For realistic models of elastic materials with W smooth no examples of absolute minimizers of I in $\mathcal{A}_p$, or even of 'strong relative minimizers' (cf Ball & Marsden [11]), that are smooth but not affine are known to the author. (The example described in §4 below could perhaps be thought of as of this type, but the properties of the stored-energy function necessary for the example have not been correlated with those of any real material, and the stress distribution is in any case trivial.) Although such smooth minimizers presumably abound their existence is hard to establish for two reasons. First, although the existence of nontrivial absolute minimizers can be proved via the direct method of the calculus of variations, no regularity theory for minimizers is available even under the mathematically most favourable realistic hypotheses; the most that is known is the partial regularity theorem of Evans [23] (see also Evans & Gariepy [24]), which still needs improvement to accommodate singular behaviour of $W(A)$ as $\det A \to 0+$. Second, although smooth non-affine solutions of the equilibrium equations (i.e. the Euler-Lagrange equations for (3.1)) are known, no extension of the field theory of the calculus of variations to dimensions $n > 1$ is available that might apply to elasticity (see Morrey [39 p15], Ball & Marsden [11]) and enable one to show that a particular solution is a minimizer.

4. MINIMIZERS SINGULAR ONLY ON THE BOUNDARY.

The work in the section is taken from Ball & James [9]. We consider a stored-energy function of a type analyzed in [5 §6.4] and Ball & Marsden [11]. To construct it let $1 < \alpha < n$, $0 < \lambda < \mu < \infty$ and let $\phi : (0,\infty) \to (0,\infty)$ satisfy (i) ϕ is smooth, (ii) $\phi' > 0$, $\phi'' > 0$, and (iii) $\phi(v) = v^{\alpha}$ for $\lambda \leqslant v \leqslant \mu$. Now choose $h : (0,\infty) \to \mathbb{R}$ satisfying (i) h is smooth, (ii) $h'' > 0$, (iii) $h(\tau) = -n\tau^{\alpha/n}$ for $\lambda^n \leqslant \tau \leqslant \mu^n$, and (iv) $h(\tau) > -n\phi(\tau^{1/n})$ for $\tau \notin [\lambda^n,\mu^n]$. Such functions h exist because $(\tau^{\alpha/n})'' < 0$ for $\lambda^n \leqslant \tau \leqslant \mu^n$.

Define

$$\Phi(v_1,\ldots,v_n) = \sum_{i=1}^{n} \phi(v_i) + h\Big(\prod_{i=1}^{n} v_i\Big), \tag{4.1}$$

and (cf (2.5))

$$W(A) = \Phi(v_1,\ldots,v_n). \tag{4.2}$$

Then W is isotropic, strictly polyconvex, and satisfies the strong ellipticity condition

$$\frac{\partial^2 W}{\partial A^i_\alpha \partial A^j_\beta}(A)\, a^i b_\alpha a^j b_\beta > 0 \quad \text{for all nonzero} \quad a,b \in \mathbb{R}^n. \tag{4.3}$$

Further, by suitably choosing Φ and h, $W(A)$ can be chosen to grow as fast as desired as $|A| \to \infty$.

It follows from the properties of Φ, h that the absolute minimizers of Φ subject to $v_i > 0$, $1 \leqslant i \leqslant n$, are given precisely by

$$v_1 = v_2 = \cdots = v_n = v \in [\lambda,\mu].$$

Consider the zero traction problem ($\partial\Omega_1$ empty) for (3.1). The absolute minimizers of I in $\mathcal{A}_p$ are those $u \in \mathcal{A}_p$ such that $Du(x)$ is a.e. an absolute minimizer of W, i.e.

$$Du(x) = v(x)Q(x) \quad \text{a.e. } x \in \Omega, \tag{4.4}$$

with $v(x) \in [\lambda,\mu]$, $Q(x) \in SO(n)$. The condition (4.4) says that u is <u>conformal</u>.

We now consider the case $n = 2$, when the conformal mappings are representable by analytic functions $w = f(z)$. Under this correspondence $v(x) = |f'(z)|$ a.e.. Thus any function f analytic in Ω and such that $0 < \inf_{z\in\Omega} |f'(z)| \leqslant \sup_{z\in\Omega} |f'(z)| < \infty$ is an absolute minimizer for some stored-energy function of the above type. To obtain a minimizer with a boundary singularity one can choose, for example, $f(z) = z^{1+i}$ (taking the principal value) with Ω the unit disc centered at $z = i$. The corresponding $u \in W^{1,\infty}$ and is invertible on $\overline{\Omega}$, but Du is discontinuous at the origin, where there is a spiral singularity. It is probably significant that the equilibrium equations, when linearized about the solution $u(x) = vx$, $v \in [\lambda,\mu]$, fail to satisfy the complementing condition of Agmon, Douglis & Nirenberg [1] with respect to the linearization of the boundary conditions (2.6). The complementing condition, which has recently been studied by Simpson & Spector [44,45] in the context of elasticity theory, is central in the study of regularity up to the boundary of linear elliptic systems.

Turning to the case $n \geq 3$, the conformal transformations are now characterized by Liouville's theorem as products of inversions. Under our regularity assumptions (i.e. $u \in W^{1,\infty}$, using (4.4)) an appropriate version of Liouville's theorem has been proved by Reshetnyak [42]. For n odd an example is given by

$$u(x) = -\frac{x}{|x|^2}. \tag{4.5}$$

If $0 \notin \overline{\Omega}$ then u satisfies (4.4) with $v(x) = |x|^{-2}$. Note that when Ω is convex this furnishes an example of a non-trivial deformation which is an absolute minimizer of the energy for a strictly polyconvex isotropic material with zero traction boundary conditions, and thus bears on a conjecture of Noll [41] (see also Truesdell [50]) to the effect that for rubber-like materials the absolute minimizer is homogeneous and unique up to rigid-body translation and rotation.

5. LIPSCHITZ MINIMIZERS.

The work in this section is also taken from Ball & James [9]. We are interested here in minimizers $u \in W^{1,\infty}$ (i.e. u Lipschitz) which are not smooth in Ω. The simplest such minimizers are piecewise affine. A piecewise affine deformation u is one for which

$$Du(x) \in \{A_1,\ldots,A_M\} \qquad \text{a.e. } x \in \Omega,$$

with $\text{meas}\, S_i > 0$ for $i = 1,\ldots,M$, where

$$S_i = \{x \in \Omega: Du(x) = A_i\},$$

$M \geq 2$ and the $A_i \in M_+^{n\times n}$ are distinct. Deformations of elastic crystals that are to a good approximation piecewise affine are commonly observed. The matrices A_i in a piecewise affine deformation are not arbitrary, and in this direction we record without proof two results. First, if $M = 2$ then necessarily

$$A_2 - A_1 = \lambda \otimes \mu \tag{5.1}$$

for some $\lambda, \mu \in \mathbb{R}^n$. This is a generalization of the jump condition of Hadamard [31]. Second, there is an example with $n = 2$ and $M = 5$ such that $\text{rank}(A_1 - A_j) = 2$ for $j = 2,\ldots,5$. In this example Du takes the value A_1 for $x^2 > 0$, while the values $A_2,\ldots,A_5$ are taken for $x^2 < 0$ on similar triangular regions whose diameter tends to zero as the interface $x^2 = 0$ is approached. The construction of the example was motivated by patterns of Dauphiné domains observed in quartz by van Tendeloo, van Landuyt & Amelinckx [51].

Piecewise affine equilibrium configurations of a homogeneous material with plane interfaces are well known to be excluded by strict rank-one convexity of W, (see, for example, [3]) that is, strict convexity of the map $t \mapsto W(A + ta \otimes b)$ for all $A \in M^{n\times n}_+$ and nonzero $a,b \in \mathbb{R}^n$. Strict rank-one convexity is implied by the strong ellipticity condition (4.3), and is not satisfied for elastic crystals by virtue of their invariance under lattice transformations (Ericksen [21]).

Suppose that $A_1,\ldots,A_M$ are each absolute minimizers of $W(x,\cdot)$ for all $x \in \Omega$. Then any piecewise affine deformation in which Du almost everywhere takes only these M values is clearly an absolute minimizer in $\mathcal{A}_p$ of I, given by (2.1), for the zero traction problem. Conversely, if $A_1,\ldots,A_N$ were for a.e. $x \in \Omega$ the only absolute minimizers of $W(x,\cdot)$ then every absolute minimizer for the zero traction problem would be piecewise affine with $1 \leqslant M \leqslant N$; however, this is impossible in elasticity because of (2.3). For this reason the following example does not apply directly to elasticity.

We consider the zero traction problem for the integral

$$I(u) = \int_\Omega W(x^3, Du(x))\,dx, \tag{5.2}$$

where $\Omega \subset \mathbb{R}^3$ contains the origin, and where for $x^3 > 0$ $W(x^3,\cdot)$ has precisely two absolute minimizers A_1, A_2 while for $x^3 < 0$ $W(x^3,\cdot)$ has the unique absolute minimizer A_3. We take $A_1 = 1 + a \otimes b$, $A_2 = 1 - a \otimes b$ and $A_3 = 1 + c \otimes e_3$, where $e_3 = (0,0,1)$ and $a,b,c \in \mathbb{R}^3$ are nonzero with b not parallel to e_3. It is easy to construct sequences $u^{(j)} \in \mathcal{A}_\infty$ for which $Du^{(j)} = A_3$ for $x^3 < 0$, $Du^{(j)}$ takes the values A_1 and A_2 for $x^3 > 1/j$ alternately on strips perpendicular to b and of width $1/j$, and such that

$$\lim_{j\to\infty} I(u^{(j)}) = \int_{\Omega\cap\{x^3<0\}} W(x^3,A_3)\,dx + \int_{\Omega\cap\{x^3>0\}} W(x^3,A_1)\,dx. \tag{5.3}$$

Hence $\inf_{\mathcal{A}_p} I$ is given by the right-hand side of (5.3), and such sequences $u^{(j)}$ are minimizing. The finely layered deformations in these minimizing sequences are similar to those observed, for example, by Saburi & Wayman [43] in shape memory martensites and by Burkart & Read [15] for indium-thallium twins. They are also reminiscent of layering in the theory of optimal design of composite materials (cf Francfort & Murat [25], Kohn & Strang [34], Lurie & Cherkaev [35], Milton [37]). We claim that the absolute minimum of I in $\mathcal{A}_p$ is not attained. If it were then any minimizer u would satisfy in a neighbourhood of the origin

$$u(x) = \begin{cases} u(0) + x + \lambda(b\cdot x)a & \text{for } x^3 > 0 \\ u(0) + x + x^3 c & \text{for } x^3 < 0, \end{cases} \tag{5.4}$$

where $\lambda \in W^{1,\infty}(\mathbb{R})$ with $\lambda(0) = 0$ and $\lambda'(t) = \pm 1$ a.e.. Applying the continuity of u on $x^3 = 0$ gives a contradiction.

6. CONTINUOUS MINIMIZERS WITH UNBOUNDED DERIVATIVES.

It is not known whether continuous minimizers u with Du unbounded can occur for realistic models of elastic materials. If such minimizers exist, they would be of importance as a mechanism for the initiation of fracture or dislocations. The existence of such singular minimizers is suggested by the one-dimensional examples of Ball & Mizel [12,13] of regular (i.e. elliptic) integrals whose minimizers among appropriate classes of absolutely continuous functions have unbounded derivatives, and do not satisfy the usual weak form of the Euler-Lagrange equation. One of these examples is the integral

$$I(u) = \int_{-1}^{1} [(u^6 - x^4)^2 |u'|^s + \varepsilon(u')^2]\,dx, \tag{6.1}$$

where $s > 27$ and $\varepsilon > 0$ is sufficiently small. The absolute minimum of I in $\mathcal{A} = \{u \in W^{1,1}(-1,1) : u(-1) = -1, u(1) = 1\}$ is attained, and every minimizer u_0 satisfies

$u_o(x) \sim |x|^{2/3} \operatorname{sign} x$, $u_o'(x) \to +\infty$ as $x \to 0$ with $u_o \in C^\infty([-1,0) \cup (0,1])$. Furthermore, if $3 \leq q \leq \infty$ then

$$\inf_{\mathcal{A} \cap W^{1,q}(-1,1)} I > \inf_{\mathcal{A}} I \quad \text{(the \underline{Lavrentiev phenomenon})}, \tag{6.2}$$

while if $\{v_m\} \subset \mathcal{A} \cap W^{1,q}(-1,1)$ with $v_m(x) \to u_o(x)$ as $m \to \infty$ for each x in some set containing arbitrarily small positive and negative numbers then

$$\lim_{m\to\infty} I(v_m) = \infty. \tag{6.3}$$

For this problem the Tonelli set $E = \{x: |u_o'(x)| = \infty\}$ consists of the single point zero, but other examples are known (Ball & Mizel [13], Davie [17]) where E is an arbitrary preassigned closed set of measure zero; these examples demonstrate the optimality of the partial regularity theorem of Tonelli [49].

7. MINIMIZERS DISCONTINUOUS AT A POINT OR LINE.

Let $n = 3$, $\Omega = \{x \in \mathbb{R}^3: |x| < 1\}$ and consider the problem of minimizing

$$I(u) = \int_\Omega W(Du(x))\,dx$$

in

$$\mathcal{A}_p = \{u \in W^{1,p}: u\big|_{\partial\Omega} = \lambda x\},$$

where $\lambda > 0$ is given. If W is $W^{1,p}$-quasiconvex then (see §3) the deformation $u(x) \equiv \lambda x$ is a global minimizer of I in $\mathcal{A}_p$. As an example, take for W the isotropic function

$$\begin{aligned} W(A) &= \mu \operatorname{tr}(A^T A) + h(\det A) \\ &= \mu(v_1^2 + v_2^2 + v_3^2) + h(v_1 v_2 v_3), \end{aligned} \tag{7.1}$$

where $\mu > 0$ and $h: (0,\infty) \to \mathbb{R}$ is a C^1 strictly convex function satisfying $\lim_{\delta \to 0+} h(\delta) = \lim_{\delta\to\infty} \frac{h(\delta)}{\delta} = \infty$. (This stored-energy function is a compressible version of the incompressible Neo-Hookean model of natural rubber.) Then W is strictly polyconvex and strongly elliptic. Further, W is $W^{1,p}$-

quasiconvex at every $A \in M_+^{3\times 3}$ if and only if $3 \leqslant p \leqslant \infty$. For $1 \leqslant p < 3$ the situation is different, as can be shown by considering radial deformations, that is deformations of the form

$$u(x) = \frac{r(R)}{R} x , \tag{7.2}$$

where $R = |x|$ and $r: [0,1] \to [0,\infty)$ is increasing. Denoting by $\mathcal{A}_p^{rad}$ the set of such radial deformations in $\mathcal{A}_p$, it can be shown [5] that there exists a number $\lambda_{cr} > 0$ such that if $0 < \lambda \leqslant \lambda_{cr}$ then $u(x) \equiv \lambda x$ is the unique absolute minimizer of I in $\mathcal{A}_p^{rad}$ for $1 \leqslant p \leqslant \infty$, but that if $\lambda > \lambda_{cr}$ then there exists a function $r_\lambda(\cdot)$ with $r_\lambda(0) > 0$ such that the corresponding u given by (7.2) is the unique absolute minimizer of I in $\mathcal{A}_p^{rad}$ for $1 \leqslant p < 3$. The nontrivial minimizers in $\mathcal{A}_p^{rad}$ for $\lambda > \lambda_{cr}$ and $1 \leqslant p < 3$ have a point discontinuity at the origin corresponding to the formation of a hole, or cavity, of radius $r_\lambda(0)$. This is a multi-dimensional example of the Lavrentiev phenomenon (see (6.2)). For developments of these results see Sivaloganathan [46,47] and Stuart [48]. (An analysis in two dimensions along similar lines gives rise to the existence of cavitating radial minimizers for appropriate strictly polyconvex stored-energy functions; when considered as plane strain deformations of three-dimensional bodies these minimizers have a line discontinuity.) It is not known whether for $1 \leqslant p < 3$ the absolute minimum of I in $\mathcal{A}_p$ is attained, and if so whether there exists a radial minimizer. If for some λ the minimum is attained by some $u \not\equiv \lambda x$ then by rescaling and patching together u one can construct infinitely many such minimizers, with more and more finely distributed patterns of holes (see Ball [7], Ball & Murat [14]).

Cavitation is a well known fracture mechanism in both polymers (Gent & Lindley [27], Denecour & Gent [19], Gent & Park [28]) and metals (Hancock & Cowling [32], Needham, Wheatley & Greenwood [40]).

There is a large literature concerning examples of singular minimizers in the multi-dimensional calculus of variations, inspired by the examples of de Giorgi [18], Giusti &

Miranda [30] and Maz'ya [36]. For a survey see Chapter II of Giaquinta [29]. It should be noted, however, that in these examples the integrand is typically convex in Du, a condition well-known to be unacceptable in elasticity.

8. CODIMENSION 1 DISCONTINUITIES.

The example of cavitation described in the preceding section demonstrates the capacity of the calculus of variations to model some types of fracture, and in particular the initiation of fracture in an ideal body without pre-existing flaws. However, the most commonly observed type of fracture, in which the deformation is discontinuous across a planar surface, cannot be treated so easily. Firstly, no such deformation can belong to the Sobolev space $W^{1,1}$, so that the meaning to be attached both to the deformation gradient Du and to the integral in (2.1) is open to discussion. Secondly, it is essential to include surface contributions to the energy, and it is not immediately obvious how this should be done because, for example, the regularity of the fracture surface is not known *a priori*. An approach of this type to fracture mechanics would be novel, and represents an interesting challenge to experts in the calculus of variations and geometric measure theory.

Related difficulties arise in analyzing the codimension 1 discontinuities occurring in dislocation theory; of particular interest here is the rich geometrical structure in the observed patterns of dislocation lines (see Friedel [26], Hull [33]).

REFERENCES

1. Agmon, S., A. Douglis & L. Nirenberg, Estimates near the boundary for solutions of elliptic partial differential equations satisfying general boundary conditions II, Comm. Pure Appl. Math. 17 (1964), 35-92.

2. Ball, J.M., Convexity conditions and existence theorems in nonlinear elasticity, Arch. Rational Mech. Anal. 63 (1977), 337-403.

3. Ball, J. M., Strict convexity, strong ellipticity, and regularity in the calculus of variations, Math. Proc. Cambridge Philos. Soc. 87 (1980), 501-513.

4. ___________ Global invertibility of Sobolev functions and the interpenetration of matter, Proc. Roy. Soc. Edinburgh Sect. A 88 (1981), 315-328.

5. ___________ Discontinuous equilibrium sólutions and cavitation in nonlinear elasticity, Philos. Trans. Roy. Soc. London Ser. A 306 (1982), 557-611.

6. ___________ Differentiability properties of symmetric and isotropic functions, Duke Math. J. 51 (1984), 699-728.

7. ___________ Minimizing sequences in thermomechanics, Proc. conference on 'Finite thermoelasticity', Rome, June 1985, to appear.

8. ___________ Does rank-one convexity imply quasiconvexity?, Proc. of workshop on 'Metastability and partial differential equations', Institute for Mathematics and its Applications, University of Minnesota, May 1985, to appear.

9. Ball, J. M. and R. D. James, in preparation.

10. Ball, J. M. and G. Knowles, Lyapunov functions for thermomechanics with spatially varying boundary temperatures, Arch. Rational Mech. Anal., to appear.

11. Ball, J. M. and J. E. Marsden, Quasiconvexity at the boundary, positivity of the second variation, and elastic stability, Arch. Rational Mech. Anal. 86 (1984), 251-277.

12. Ball, J. M. and V. J. Mizel, Singular minimizers for regular one-dimensional problems in the calculus of variations, Bull. Amer. Math. Soc. 11 (1984), 143-146.

13. Ball. J. M. and V. J. Mizel, One-dimensional variational problems whose minimizers do not satisfy the Euler-Lagrange equation, Arch. Rational Mech. Anal. 90 (1985), 325-388.

14. Ball, J. M. and F. Murat, $W^{1,p}$-quasiconvexity and variational problems for multiple integrals, J. Funct. Anal. 58 (1984), 225-253.

15. Burkart, M. W. and T. A. Read, Diffusionless phase change in the Indium-Thallium system, J. Metals (1953), 1516-1524.

16. Ciarlet, P. G. and J. Nečas, Injectivité presque partout, auto-contact, et non-interpénétrabilité en elasticité non linéaire tridimensionelle, C.R. Acad. Sc. Paris 301 I (1985), 621-624.

17. Davie, A. M., to appear.

18. de Giorgi, E., Un esempio di estremali discontinue per un problema variazionale di tipo ellittico, Boll. Un. Mat. Ital. (4) 1, (1968), 135-137.

19. Denecour, R. L. and A. N. Gent, Bubble formation in vulcanized rubbers, J. Polymer Science A-2 6 (1968), 1853-1861.

20. Duhem, P., Traité d'energetique ou de thermodynamique générale, Gauthier-Villars, Paris, 1911.

21. Ericksen, J. L., Thermoelastic stability, Proc. 5th National Cong. Appl. Mech. (1966), 187-193.

22. ______________ Special topics in elastostatics, in Advances in Applied Mechanics vol. 17, Academic Press, New York, 1977.

23. Evans, L. C., Quasiconvexity and partial regularity in the calculus of variations, to appear.

24. Evans, L. C. and R. F. Gariepy, Blow-up, compactness and partial regularity in the calculus of variations, to appear.

25. Francfort, G. A. and F. Murat, Homogenization and optimal bounds in linear elasticity, Arch. Rational Mech. Anal., to appear.

26. Friedel, J., Dislocations, Pergamon, Oxford, 1964.

27. Gent, A. N. and P. B. Lindley, Internal rupture of bonded rubber cylinders in tension, Proc. Roy. Soc. London Ser.A 249 (1958), 195-205.

28. Gent, A. N. and B. Park, Failure processes in elastomers at or near a rigid spherical inclusion, J. Materials Science 19 (1984), 1947-1956.

29. Giaquinta, M., Multiple integrals in the calculus of variations and nonlinear elliptic systems, Princeton University Press, Princeton, New Jersey, 1983.

30. Giusti, E. and M. Miranda, Un esempio di soluzioni discontinue per un problema di minimo relativo ad un integrale regolare del calcolo delle variazioni, Boll. Un. Mat. Ital. (4) 1 (1968), 219-225.

31. Hadamard, J., Leçons sur la propagation des ondes, Hermann, Paris, 1903.

32. Hancock, J. W. and M. J. Cowling, Role of state of stress in crack-tip failure processes, Metal Science (1980), 293-304.

33. Hull, D., Introduction to dislocations, Pergamon, Oxford, 1965.

34. Kohn, R. V. and G. Strang, Optimal design and relaxation of variational problems, Comm. Pure Appl. Math., to appear.

35. Lurie, K. A. and A. V. Cherkaev, Exact estimates of conductivity of composites formed by two isotropically conducting media taken in prescribed proportion, Proc. Roy. Soc. Edinburgh Sect. A 99 (1984), 71-88.

36. Maz'ya, V. G., Examples of nonregular solutions of quasilinear elliptic equations with analytic coefficients, Funktsional. Anal. i Prilozhen 2 (1968), 53-57.

37. Milton, G. W., Modelling the properties of composites by laminates, in Proc. of the 'Workshop on homogenization and effective moduli of materials and media', Minneapolis, 1984, to appear.

38. Morrey, C. B., Quasi-convexity and the lower semicontinuity of multiple integrals, Pacific J. Math. 2 (1952), 25-53.

39. ___________ Multiple integrals in the calculus of variations, Springer, Berlin, 1966.

40. Needham, N. G., J. E. Wheatley and G. W. Greenwood, The creep fracture of copper and magnesium, Acta Metallurgica 23 (1974), 23-27.

41. Noll, W., A general framework for problems in the statics of finite elasticity in Contemporary Development in Continuum Mechanics and Partial Differential Equations, North-Holland, Amsterdam, 1978, pp363-387.

42. Reshetnyak, Yu. G., On the stability of conformal mappings in multidimensional spaces, Siberian Math. J. 8 (1967), 69-85.

43. Saburi, T. and C. M. Wayman, Crystallographic similarities in shape memory martensites, Acta Metallurgica 27 (1979), 979-995.

44. Simpson, H. C. and S. J. Spector, On failure of the complementing condition and nonuniqueness in linear elastostatics, J. Elasticity 15 (1985), 229-231.

45. ______________________ On Hadamard stability in finite elasticity, to appear.

46. Sivaloganathan, J., Uniqueness of regular and singular equilibria in radial elasticity, Arch. Rational Mech. Anal., to appear.

47. ____________ A field theory approach to stability of radial equilibria in nonlinear elasticity, Math. Proc. Cambridge Philos. Soc., to appear.

48. Stuart, C. A., Radially symmetric cavitation for hyperelastic materials, Ann. Inst. H. Poincaré - Analyse non linéaire 2 (1985), 1-20.

49. Tonelli, L., Fondamenti di calcolo delle variazioni, Nicola Zanichelli Editore, Bologna, 1921.

50. Truesdell, C., Some challenges to analysis by rational thermomechanics, in Contemporary Developments in Continuum Mechanics and Partial Differential Equations, North-Holland, Amsterdam, 1978, pp495-603.

51. Tendeloo, G. van, J. van Landuyt and S. Amelinckx, The $\alpha \rightarrow \beta$ phase transition in quartz and $ALPO_4$ as studied by electron microscopy and diffraction, Phys. Status Solidi(A) 33 (1976), 723-735.

Acknowledgements

I would like to thank the organizers, and other members of the Mathematics Research Center, for enabling me to participate in a stimulating and enjoyable conference. The research reported here was partially supported by a U.K. S.E.R.C. Senior Fellowship.

Department of Mathematics
Heriot-Watt University
Riccarton
Edinburgh EH14 4AS
Scotland, U.K.

Nonlinear Elliptic Equations Involving the Critical Sobolev Exponent—Survey and Perspectives

Haïm Brezis

1. INTRODUCTION

Let Ω be a (smooth) bounded domain in $\mathbf{R}^N$ with $N \geq 3$. Consider the problem of finding a function $u(x)$ which satisfies

$$\left.\begin{aligned} -\Delta u &= u^p + a(x)u && \text{on } \Omega, \\ u &> 0 && \text{on } \Omega, \\ u &= 0 && \text{on } \partial\Omega, \end{aligned}\right\} \tag{I}$$

where $p = (N + 2)/(N - 2)$ and $a(x)$ is a given smooth function. The question is to determine conditions on $a(x)$ and Ω which guarantee that (I) has a solution. For example, an obvious <u>necessary condition</u> is that the linear operator $L = -\Delta - a$ should be <u>coercive</u>, that is,

$$\int (|\nabla\varphi|^2 - a\varphi^2) \geq \delta \int \varphi^2 \qquad \forall \varphi \in H_0^1 \quad \text{with} \quad \delta > 0 . \tag{1}$$

Indeed, let μ_1 denote the first eigenvalue of L (with zero Dirichlet condition) and let $\varphi_1 > 0$ be a corresponding eigenfunction. Multiplying (I) through by φ_1 we obtain

$$\mu_1 \int u\varphi_1 = \int u^p \varphi_1 ,$$

so that μ_1 must be positive if a solution of (I) exists. It is tempting to assert that (1) is also a sufficient condition; and this is indeed the case when $1 < p < (N + 2)/(N - 2)$.

DIRECTIONS IN PARTIAL
DIFFERENTIAL EQUATIONS

ISBN 0-12-195255-X

However, when $p = (N+2)/(N-2)$ the situation is completely different. A celebrated result of Pohozaev [34] asserts that if Ω is starshaped and $a(x) \equiv 0$ there is no solution of (I) (despite the fact that (1) holds). Therefore, extra assumptions, in addition to (1), are needed. Two different kinds of conditions have been found so far:

a) The function $a(x)$ is positive somewhere on Ω when $N \geq 4$ (and no assumption on Ω); a more delicate (and global) assumption when $N = 3$.

b) The domain Ω has nontrivial topology - for example Ω has a "hole" - (and no assumption on the function $a(x)$).

Case a) has been investigated in [17] following a technique introduced by Th. Aubin [8]; it is discussed in Sections 2 and 3. Case b) is related to the splendid work of Bahri-Coron [10-11] which is discussed in Section 5.

A closely related problem is to find a function $u(x)$ which satisfies

$$\left.\begin{aligned} -\Delta u &= |u|^{p-1}u + a(x)u && \text{on } \Omega, \\ u &\not\equiv 0 && \text{on } \Omega, \\ u &= 0 && \text{on } \partial\Omega, \end{aligned}\right\} \qquad \text{(II)}$$

where, again, $p = (N+2)/(N-2)$ and $a(x)$ is a given (smooth) function. Clearly, any solution of (I) is a solution of (II). Nevertheless, problem (II) is of independent interest for the following reasons:

a) The conditions to impose on $a(x)$ or Ω may be weaker; for example (1) is not anymore a necessary condition.

b) Problem (II) may have many solutions - possibly infinitely many - in cases where (I) has just one solution.

Problems such as (I) or (II) have attracted much attention in recent years. Despite their simple form they have a very rich structure. The main reason is that they admit a <u>variational formulation which lacks the Palais-Smale (PS) condition</u>. For example, solutions of (II) correspond to nonzero critical points of the functional

$$F(u) = \frac{1}{2}\int |\nabla u|^2 - \frac{1}{p+1}\int |u|^{p+1} - \frac{1}{2}\int au^2 .$$

The functional F is well defined on the Sobolev space H_0^1 (because of the Sobolev imbedding $H^1 \subset L^{p+1} = L^{2N/(N-2)}$. However, F does not satisfy the (PS) compactness condition (because this imbedding is not compact). So that, all the standard variational techniques (minimization, Morse theory, Ljusternik-Schnirelman theory etc. ...) do not apply directly to (I) or (II). These kinds of problems are fascinating experimental laboratories for discovering new tools to overcome this lack of compactness. Related questions - such as the Yamabe problem - occur in geometry (see e.g. Kazdan [28]) and in physics (see e.g. the work of Taubes [40] dealing with the Yang-Mills equations in dimension four).

The plan is the following:

1. Introduction.
2. Problem (I) when $N \geq 4$.
3. Problem (I) when $N = 3$.
4. Problem (II).
5. The effect of topology: the work of Bahri-Coron.
6. Various related problems.

2. PROBLEM (I) WHEN $N \geq 4$

Throughout Sections 2 and 3 we shall assume that (1) holds. The first - and simplest - approach in order to find a solution of (I) is minimization with constraint. More precisely, set

$$J = \inf_{\varphi \in H_0^1} \left\{\frac{|\nabla\varphi|^2 - a\varphi^2}{\|\varphi\|_{p+1}^2}\right\} = \inf_{\substack{\varphi \in H_0^1 \\ \|\varphi\|_{p+1}=1}} \left\{\int |\nabla\varphi|^2 - a\varphi^2\right\}. \tag{2}$$

Note that $J > 0$ (because (1) holds). It is not clear whether this infimum is achieved - the reason is that the constraint $\|\varphi\|_{p+1} = 1$ is not preserved under weak H^1 limits. However, if we assume that J is achieved, then we obtain a solution of (I). Indeed let φ be a minimizer; we may always suppose that $\varphi \geq 0$ - otherwise replace φ by $|\varphi|$. The Euler equation for (2) is

$$-\Delta\varphi - a\varphi = J\varphi^p \quad \text{on} \quad \Omega .$$

It follows from the strong maximum principle that $\varphi > 0$ on Ω and finally we obtain a solution of (I) by scaling out the constant J.

As pointed out by Th. Aubin [7-8] the best Sobolev constant S plays an important role in such problems. S is defined by

$$S = \operatorname*{Inf}_{\varphi \in H_0^1} \left\{ \frac{\int |\nabla\varphi|^2}{\|\varphi\|_{p+1}^2} \right\} = \operatorname*{Inf}_{\substack{\varphi \in H_0^1 \\ \|\varphi\|_{p+1}=1}} \left\{ \int |\nabla\varphi|^2 \right\} . \tag{3}$$

It is easy to see that S is independent of Ω (it depends only on N) and that the infimum in (3) is <u>not</u> achieved in any bounded domain.

The success (or the failure) of this minimization approach is <u>completely settled</u> by the following:

<u>Theorem 1</u> (Brezis-Nirenberg): Assume Ω is any bounded domain in $\mathbb{R}^N$ with $N > 4$. The following properties are equivalent:

$$a(x) > 0 \quad \text{somewhere on} \quad \Omega , \tag{4}$$

$$J < S , \tag{5}$$

$$J \text{ is achieved (and therefore (I) has a solution)} . \tag{6}$$

The proof of Theorem 1 is essentially contained in [17] (see also [14]). The most technical part is the proof of (4) ==> (5), which is achieved by constructing explicitly a function φ such that

$$Q(\varphi) = \frac{\int |\nabla\varphi|^2 - a\varphi^2}{\|\varphi\|_{p+1}^2} < S .$$

More precisely, φ is chosen of the form

$$\varphi(x) = \zeta(x)/(\varepsilon + |x|^2)^{(N-2)/2}$$

with $\varepsilon > 0$ small enough and $\zeta \in C_0^\infty(\Omega)$ is such that $\zeta \equiv 1$ near a point x_0 where $a(x_0) > 0$. It is in this part of the argument that the assumption $N > 4$ comes in.

In fact, the implication (4) ==> (5) fails when $N = 3$. The other implications (5) ==> (6) and (6) ==> (4) are rather easy and hold even when $N > 3$. This leads us to:

Question 1: Is there a direct proof of the implication (6) ==> (5) which does not make use of (4)? Is it true that (6) ==> (5) when $N = 3$?

There are related - "dual" - maximization problems which are likely to have solutions; for example:

Question 2: Assume $a(x) > 0$ somewhere on Ω and consider

$$\sup_{\substack{\varphi \in H_0^1 \\ \int |\nabla \varphi|^2 \leq 1}} \left\{\frac{1}{p+1} \int |\varphi|^{p+1} + \frac{1}{2} \int a\varphi^2\right\} .$$

Is this supremum achieved when $N > 4$? Same question when $N = 3$.

Note that the minimization technique (2) may not be the right approach for finding a solution of Problem (I). In particular, if $a(x) < 0$ everywhere on Ω, Problem (I) may still have a solution. Of course, Theorem 1 prevents that solution from being reached by minimization and it is a challenge to find the appropriate tool. Here are two examples:

Example 1: Assume (for simplicity) that $a(x) \equiv 0$ and that Ω has nontrivial topology. Then, Problem (I) has a solution (see Section 5).

Example 2: Let Ω be any domain (for instance a ball) and fix any function $f \in C_0^\infty(\Omega)$ with $f > 0$, $f \not\equiv 0$. Let v be the solution of the problem

$$\begin{cases} -\Delta v = f & \text{on } \Omega , \\ v = 0 & \text{on } \partial\Omega , \end{cases}$$

so that $v > 0$ on Ω. Set

$$a = \frac{f}{v} - \mu^{p-1} v^{p-1}$$

where μ is a constant large enough so that $a < 0$ on Ω. It is clear that $u = \mu v$ is a solution of Problem (I).

This leads us to the following:

Question 3: Assume $a(x) < 0$ everywhere on Ω. Find conditions on $a(x)$ which guarantee that Problem (I) has a solution. The answer is unknown even if Ω is a ball and $a(x)$ is a radial function.

Remark 1: The Pohozaev identity applied to a solution u of Problem (I) says that

$$\int_\Omega \left(a + \frac{1}{2} \Sigma x_i \frac{\partial a}{\partial x_i}\right)u^2 = \frac{1}{2} \int_{\partial\Omega} (x \cdot n)\left(\frac{\partial u}{\partial n}\right)^2 . \qquad (7)$$

If Ω is starshaped with respect to the origin, an obvious necessary condition for the existence of a solution is that $(a + \frac{1}{2} \Sigma x_i \frac{\partial a}{\partial x_i})$ should be positive somewhere on Ω. In particular, if $a(x) \equiv \lambda$ is a constant and Ω is starshaped we find that Problem (I) has a solution for every $\lambda \in (0, \lambda_1)$ and no solution otherwise, where λ_1 denotes the first eigenvalue of $-\Delta$ with zero Dirichlet condition.

3. PROBLEM (I) WHEN N = 3

For some strange reason this case is much more delicate than the case $N \geq 4$. The most recent contribution is due, independently to B. McLeod [31] and R. Schoen [37] (actually R. Schoen was working on the Yamabe problem, but his idea for solving the Yamabe problem in low dimension applies in our situation as well). In order to describe their result one has to introduce the Green's function $G(x,y)$ of the operator $L = -\Delta - a$, that is, $G(\cdot,y)$ is the solution of

$$\begin{cases} -\Delta G - aG = \delta(\cdot - y) & \text{on } \Omega \\ G = 0 & \text{on } \partial\Omega . \end{cases}$$

Note that G is well defined because of assumption (1). Write

$$G(x,y) = \frac{1}{4\pi|x - y|} + g(x,y) ;$$

the function $g(x,y)$ is called the regular part of the Green's function and g is continuous on $\Omega \times \Omega$ (including the diagonal (x,x)).

Theorem 2 (McLeod, Schoen): Assume Ω is any bounded domain in $\mathbb{R}^3$. The following implications hold:

$g(x,x) > 0$ somewhere on Ω , (8)

$J < S$, (9)

J is achieved (and therefore (I) has a solution) . (10)

For the proof of Theorem 2 we refer to [31], [37] and also [14]. As in Theorem 1, the most technical part is the proof of (8) ==> (9) which is achieved by constructing a function φ such that $Q(\varphi) < S$. As pointed out by Bahri and Coron [11], it is convenient to choose $\varphi = \varphi_\varepsilon$ to be the solution of the problem:

$$\begin{cases} -\Delta\varphi_\varepsilon - a\varphi_\varepsilon = \dfrac{1}{(\varepsilon + |x - x_0|^2)^{5/2}} & \text{on } \Omega , \\ \varphi_\varepsilon = 0 & \text{on } \partial\Omega \end{cases}$$

where x_0 is any point in Ω such that $g(x_0,x_0) > 0$. An easy expansion as $\varepsilon \to 0$ (see [14]) shows that

$$Q(\varphi_\varepsilon) = S - Cg(x_0,x_0)\varepsilon^{1/2} + o(\varepsilon^{1/2})$$

where C is some positive (universal) constant.

It is tempting to compare Theorems 1 and 2 and to raise the following:

<u>Question 4</u>: Are the properties (8), (9), (10) equivalent?

The answer to Question 4 is not known, even in the case where $a(x) \equiv \lambda$ is a constant and Ω is, say, convex. However, the case where $a(x) \equiv \lambda$ is a constant and Ω is a ball is completely settled (see [17]): Problem (I) has a solution if $\lambda \in (\frac{1}{4}\lambda_1, \lambda_1)$ and no solution otherwise. In that case the answer to Question 4 is positive.

As in Section 2, I must emphasize that the minimization technique (2) may not be the right approach for finding a solution of Problem (I). In particular, if $a(x) < 0$ everywhere on Ω, it follows (from the maximum principle) that $g(x,x) < 0$ everywhere on Ω - and Problem (I) may still have a solution (see Example 2 in Section 2). Again, it is a challenge to find other tools besides minimization.

4. <u>PROBLEM (II)</u>

Here, assumption (1) plays no role and is not required. When p is <u>subcritical</u>, i.e. $1 < p < (N + 2)/(N - 2)$, it is well-known (see [1]) that Problem (II) possesses infinitely many solutions without any additional assumption (on Ω or $a(x)$). However, in the critical case, $p = (N + 2)/(N - 2)$ further assumptions are needed. Indeed, the Pohozaev identity (7) still holds for any solution u of Problem (II). In particular, if Ω is starshaped and $a(x) \equiv \lambda$ is a constant, we find that λ must be positive if a solution of (II) exists. In view of this obstruction one can anticipate two kinds of assumptions which may force solutions to exist:

a) the function $a(x)$ is positive somewhere on Ω (and no assumption about Ω);

b) the domain Ω has nontrivial topology (and no assumption about $a(x)$).

So far, little is known about Problem (II), but some interesting progresses have been achieved during the last two years for the case $a(x) \equiv \lambda$ is a constant:

Theorem 3 (Capozzi-Fortunato-Palmieri): Assume Ω is any bounded domain in $\mathbf{R}^N$ with $N \geq 4$. Then, for every $\lambda > 0$, Problem (II) has a solution.

For the proof we refer to [18]. An elegant alternative proof has been given by Ambrosetti-Struwe [3], using a dual formulation proposed in [2] and which is similar to the dual formulation of Clarke-Ekeland [23] for Hamiltonian systems (see also [13]).

Theorem 4 (Cerami-Solimini-Struwe, D. Zhang): Assume Ω is any bounded domain in $\mathbf{R}^N$ with $N \geq 6$. Then, for every $\lambda \in (0,\lambda_1)$, Problem (II) has at least two pairs of solutions.

Theorem 5 (Solimini): Assume Ω is a ball in $\mathbf{R}^N$ with $N \geq 7$. Then for every $\lambda \in (0,\infty)$, Problem (II) has infinitely many radial solutions.

For the proof we refer to [38] (see also [21], [27] and [42] for related results). The idea is to use a min-max argument as in Ambrosetti-Rabinowitz [1] - except that the failure of the (PS) condition requires additional devices inspired by [17] and [39]. A result of a more "local" nature (for $\lambda < \lambda_j$ and $\lambda_j - \lambda$ small enough) had been obtained previously by Cerami-Fortunato-Struwe [20]. Naturally, we are led to the following:

Question 5: Assume Ω is any bounded domain in $\mathbf{R}^N$ with $N > 4$ and assume $a(x) > 0$ somewhere in Ω. Does Problem (II) have (at least) one solution? infinitely many solutions? What happens when $a(x) < 0$ everywhere on Ω? How about $N = 3$?

5. THE EFFECT OF TOPOLOGY: THE WORK OF BAHRI-CORON.

A remarkable result of Bahri-Coron [11] shows that the topology of Ω plays an important role which may cancel the "Pohozaev obstruction":

Theorem 6: Assume Ω is a domain in $\mathbf{R}^N$, $N > 3$, with nontrivial topology and assume $a(x) \equiv 0$. Then, there exists a solution of Problem (I).

The precise meaning of the assumption "Ω has nontrivial topology" is expressed in terms of homology groups: there exists an integer $k > 1$ such that either $H_{2k-1}(\Omega;\mathbf{Q}) \neq 0$ or $H_k(\Omega;\mathbf{Z}/2\mathbf{Z}) \neq 0$. When $N = 3$, Ω has nontrivial topology iff Ω is not contractible. When $N > 4$, the assumption "Ω has nontrivial topology" covers a large variety of domains. For example, a domain which has the topology of a solid torus or any domain with a hole is OK. (However, when $N > 4$ it is not known whether the conclusion of Theorem 6 holds under the only assumption that Ω is not contractible.)

The result of Bahri-Coron prompts many questions:

Question 6: Assume Ω has nontrivial topology and $a(x)$ is any function such that (1) holds. Is there always a solution of Problem (I)? Likewise, if Ω is replaced by a manifold M of dimension N, without boundary.

Question 7 (Bott): Does the topology of Ω affect the number of solutions of Problem (I)? For example, if Ω has two holes, are there at least two solutions of Problem (I)?

Question 8: Assume Ω has nontrivial topology and $a(x)$ is any function (without assuming (1)). Is there at least one solution of Problem (II)? Are there infinitely many solutions of Problem (II)? Likewise, if Ω is replaced by a manifold M without boundary.

The proof of Theorem 6 is rather difficult and it involves many new ideas. I will sketch briefly the basic structure of the argument. First, recall that the minimization approach used in Sections 2 and 3 definitely fails. Indeed, if we consider

$$\operatorname*{Inf}_{\substack{\varphi \in H_0^1 \\ \|\varphi\|_{p+1}=1}} \{\int |\nabla\varphi|^2\} ,$$

this is precisely the best Sobolev constant S and it is never achieved in any domain. Therefore, if Problem (I) has a solution it corresponds, in the variational formulation, to a critical point which is not a minimum. That is why it is tempting to use Morse theory. Unfortunately, the classical Morse theory involves a compactness condition - namely the (PS) condition which fails in our problem. What Bahri and Coron have done is to analyze very precisely how the (PS) condition fails and they have overcome this lack of compactness by a kind of "compactification at infinity". The abstract setting is the following:

Let H be a Hilbert space. Let $F : H \to \mathbf{R}$ be a function of class C^2. Given $a \in \mathbf{R}$ we set, as usual

$$F_a = \{u \in H;\ F(u) < a\} .$$

A critical point of F is an element $u \in H$ such that $F'(u) = 0$. A critical value c is a real number such that $c = F(u)$ for some critical point u. The standard (PS) condition says that:

$$\left.\begin{array}{l}\text{every sequence } (u_n) \text{ in } H \text{ such that } |F(u_n)| \\ \text{is bounded and } \|F'(u_n)\| \to 0 \text{ is relatively} \\ \text{compact in } H .\end{array}\right\} \quad \text{(PS)}$$

For our purpose, it is convenient to use a weaker, "localized", version of the (PS) condition which has been introduced in [16]. Given $c \in \mathbf{R}$ we say that $(PS)_c$ holds if

$$\left.\begin{array}{l}\text{every sequence } (u_n) \text{ in } H \text{ such that}\\ F(u_n) \to c \text{ and } \|F'(u_n)\| \to 0 \text{ is}\\ \text{relatively compact in } H .\end{array}\right\} \qquad (PS)_c$$

The (PS) condition prevents critical points from "leaking at infinity". More precisely, if the $(PS)_c$ conditions fails at some level c, we may say that c is a critical value which corresponds to a "critical point at infinity" - a concept introduced by A. Bahri in [9].

A basic principle of Morse theory - based on a standard deformation argument along the gradient flow - asserts that $F_a \simeq F_b$ (homotopy equivalence) provided:

i) F has no critical value in the interval $[a,b]$,

ii) F satisfies $(PS)_c$ for every $c \in [a,b]$.

For the study of Problem (I), we take $H = H_0^1(\Omega)$ and

$$F(u) = \frac{1}{2} \int |\nabla u|^2 - \frac{1}{p+1} \int |u^+|^{p+1}$$

where $u^+ = \max(u,0)$. Clearly $u = 0$ is a critical point of F and the nonzero critical points of F correspond precisely to the solutions of Problem (I). The proof of Theorem 6 is by contradiction and henceforth we shall assume that

$$u = 0 \text{ is the only critical point of } F . \qquad (11)$$

It is rather easy to see that the $(PS)_c$ condition fails at the levels $c = k\Sigma$ where $k = 1,2,3,\ldots$ and $\Sigma = (1/N)S^{N/2}$. This is done by constructing explicitly a sequence (u_n) such that $u_n \rightharpoonup 0$ weakly in H_0^1 and $F(u_n) \to k\Sigma$ (see e.g. [14]). It is a striking fact - which is much more difficult to prove - that the $(PS)_c$ condition fails only at the levels $k\Sigma$; the argument relies on a blowup technique originally due to J. Sacks-Uhlenbeck [36] and subsequently developed by many authors (e.g. P. L. Lions [30], M. Struwe [39], H. Brezis-J. M. Coron [15]); for details, see [11] and [14].

In view of assumption (11) and the basic principle of Morse theory we see that changes in the topology of the sets F_c may occur only across the levels $c = k\Sigma$, $k = 0,1,2,\ldots$. Bahri and Coron have obtained an explicit representation of the jump in topology at each of the levels $k\Sigma$. For example, the pair $(F_{\Sigma+\varepsilon}, F_{\Sigma-\varepsilon})$ has the same topology (homotopy equivalence) as the pair $(\Omega \times B^1, \Omega \times S^0)$ where $B^1 = \{x \in \mathbf{R};\ |x| \leq 1\}$ and $S^0 = \partial B^1 = \{-1,+1\}$. This is achieved by a deformation method, i.e. pushing down with the gradient flow. But it requires a delicate Morse analysis of the Hessian matrix D^2F near the critical points infinity - which correspond to sequences (u_n) such that $F(u_n) \to \Sigma$ and $F'(u_n) \to 0$.

At the level $k\Sigma$, $k \geq 2$, there is still an explicit formula describing the topology of the pair $(F_{k\Sigma+\varepsilon}, F_{k\Sigma-\varepsilon})$ in terms of Ω and the Green's function $G(x,y)$ of Δ (with zero Dirichlet condition). However, the formula is complicated and I shall not write it down; see e.g. [11] and [14].

Using that formula, Bahri and Coron are able to see that:

1) For any domain Ω, there is an integer k_0 (depending on Ω) such that:

   $(F_{k\Sigma+\varepsilon}, F_{k\Sigma-\varepsilon})$ is trivial for $k \geq k_0$;

   in other words $F_{k\Sigma+\varepsilon} \simeq F_{k\Sigma-\varepsilon}$ for $k \geq k_0$, i.e. there is no change in topology in F_c for $c \geq k_0\Sigma$.

2) Assuming that Ω has nontrivial topology, then for each $k \geq 1$ the pair $(F_{k\Sigma+\varepsilon}, F_{k\Sigma-\varepsilon})$ is nontrivial.

This last claim is by induction on k (it is obvious when $k = 1$) and relies heavily on tools from algebraic topology. Therefore, we are led to a contradiction and, thus assumption (11) is absurd.

6. VARIOUS RELATED PROBLEMS

6.1. The equation $-\Delta u = u^p + \mu u^q$.

Let Ω be a (smooth) bounded domain in $\mathbf{R}^N$ with $N \geq 3$. Consider the problem of finding a function $u(x)$ which satisfies

$$\left.\begin{aligned} -\Delta u &= u^p + \mu u^q && \text{on } \Omega \\ u &> 0 && \text{on } \Omega \\ u &= 0 && \text{on } \partial\Omega \end{aligned}\right\} \qquad (12)$$

where $p = (N + 2)/(N - 2)$, $1 < q < p$, and μ is a constant.

Theorem 7 (Brezis-Nirenberg): Let Ω be any domain in $\mathbf{R}^N$ with $N \geq 4$. Then, for every $\mu > 0$ there exists a solution of (12).

For the proof we refer to [17]. Again, the argument consists of finding a nonzero critical point of the functional

$$F(u) = \frac{1}{2} \int |\nabla u|^2 - \frac{1}{p+1} \int (u^+)^{p+1} - \frac{\mu}{q+1} \int (u^+)^{q+1} .$$

The mountain pass theorem of Ambrosetti-Rabinowitz [1] applies, except for the (PS) condition and therefore an additional argument is needed. Roughly speaking, we are saved by the fact that F satisfies the $(PS)_c$ condition provided $c < \Sigma$ (the same Σ as in Section 5).

If Ω is starshaped, Pohozaev's identity shows that there is no solution of (12) when $\mu \leq 0$. However, the results of Section 5 suggest the following:

Question 9. Assume $\Omega \subset \mathbf{R}^N$, with $N \geq 3$, has nontrivial topology and $\mu \in \mathbf{R}$. Is there a solution of Problem (12)? Likewise on a manifold M without boundary.

When $N = 3$, so that $p = 5$, the situation is rather peculiar. The following results have been established:

Theorem 8 (Brezis-Nirenberg): Assume Ω is any domain in $\mathbf{R}^3$ and $3 < q < 5$. Then, for every $\mu > 0$ there exists a solution of (12). However, if $1 < q \leq 3$ one has to assume that μ is large enough in order to have a solution; if $1 < q \leq 3$ and Ω is starshaped no solution exists for $\mu \geq 0$ small.

For the proof, see [17]. There is a more precise result in the case of a ball:

Theorem 9 (Atkinson-Peletier): Assume Ω is a ball in $\mathbf{R}^3$ and $1 < q < 3$. Then, there exists a constant $\mu_0 > 0$ such that:

(i) if $\mu > \mu_0$, Problem (12) has at least two solutions,

(ii) if $\mu < \mu_0$, Problem (12) has no solution.

For the proof see [4]. Naturally, we may ask the following:

Question 10: Assume Ω is any domain in $\mathbf{R}^3$ and $1 < q < 3$. Does Problem (12) have at least two solutions for μ large enough?

Of course, we may also replace (12) by

$$\left.\begin{aligned} -\Delta u &= |u|^{p-1}u + \mu|u|^{q-1}u && \text{on } \Omega \\ u &\not\equiv 0 && \text{on } \Omega \\ u &= 0 && \text{on } \partial\Omega \end{aligned}\right\} \qquad (13)$$

and ask:

Question 11: Are there infinitely many solutions of Problem (13) under the natural restrictions? By the natural restrictions we mean one of the following cases:

a) $N \geq 4$, Ω is any domain and $\mu > 0$,

b) $N = 3$, Ω is any domain and $3 < q < 5$ or $1 < q \leq 3$ and μ is large enough,

c) $N \geq 3$, Ω is a domain with topology and no condition on $\mu \in \mathbf{R}$.

Along these lines, I should also mention a work by F. Atkinson-L. Peletier-J. Serrin (in preparation) dealing with the problem:

$$\left.\begin{aligned} -\operatorname{div}(|\nabla u|^{m-2}\nabla u) &= u^p + \mu u^q && \text{on } \Omega, \\ u &> 0 && \text{on } \Omega, \\ u &= 0 && \text{on } \partial\Omega, \end{aligned}\right\} \qquad (14)$$

in the critical exponent case $\left(\frac{1}{p+1} = \frac{1}{m} - \frac{1}{N}\right)$ and $1 < q < p$ (see also [35]).

6.2. The equation $-\Delta u - \frac{1}{2} x \cdot \nabla u = u^p + \lambda u$ on $\mathbf{R}^N$.

Consider the problem of finding a function $u(x)$ which satisfies

$$\left.\begin{aligned} -\Delta u - \tfrac{1}{2} x \cdot \nabla u &= u^p + \lambda u && \text{on } \mathbf{R}^N, \\ u &> 0 && \text{on } \mathbf{R}^N, \\ \int (|u|^2 + |\nabla u|^2)\exp(|x|^2/4) &< \infty . \end{aligned}\right\} \qquad (15)$$

Such a problem arises in the study of self-similar solutions of the evolution equation $u_t - \Delta u = u^p$. Again, the critical exponent $p = (N + 2)/(N - 2)$ plays a special role and one has the following:

Theorem 10 (Escobedo-Kavian, Atkinson-Peletier): Assume $p = (N + 2)/(N - 2)$. Then

(i) if $N \geq 4$ and $\lambda \in (\frac{N}{4}, \frac{N}{2})$ there is at least one solution of (14); no solution when $\lambda \notin (\frac{N}{4}, \frac{N}{2})$,

(ii) if $N = 3$ and $\lambda \in (1, \frac{3}{2})$ there is at least one solution of (14); no solution when $\lambda \notin (1, \frac{3}{2})$.

For the proofs, see [26] and [5]. There is a striking similarity with the results of Section 2 (and 3). For example, when $N \geq 4$, the restriction $\lambda > N/4$ corresponds to a "Pohozaev type obstruction" while the restriction $\lambda < N/2$ corresponds to the assumption that the linear operator $L = -\Delta - \frac{1}{2} x \cdot \nabla - \lambda$ is coercive on an appropriate weighted Sobolev space.

6.3. The equation $-\Delta u = K(x)u^p + a(x)u$.

Consider the problem of finding a function $u(x)$ which satisfies

$$\left.\begin{aligned} Lu \equiv -\Delta u - a(x)u &= K(x)u^p && \text{on } \Omega \\ u &> 0 && \text{on } \Omega \\ u &= 0 && \text{on } \partial\Omega \end{aligned}\right\} \tag{16}$$

where $p = (N + 2)/(N - 2)$, $a(x)$ and $K(x)$ are given (smooth) functions. Likewise, one may consider the same problem on a manifold M without boundary - which is of interest in differential geometry see e.g. [28].

Little is known about the solvability (or nonsolvability) of (15). Obstructions to existence (related to a Pohozaev-type identity) have been found by Kazdan-Warner [29] and also Bourguignon-Ezin [12]. Some positive results have been obtained by Ni [33] (when $\Omega = \mathbf{R}^N$), Escobar-Schoen [25] and Escobar [24]. The most interesting positive result is the following:

Theorem 11 (Bahri-Coron): Assume $M = S^3$ and L is the conformal Laplacian on S^3. Let $K(x)$ be a positive function of class C^2 with finitely many nondegenerate critical points $y_1, y_2, \ldots, y_n$, such that $(LK)(y_i) \neq 0$. Let k_i be the Morse index of $K(x)$ at y_i and assume that

$$\sum_{(LK)(y_i)<0} (-1)^{k_i} \neq -1 .$$

Then, there exists a solution of (15) on S^3.

The proof (see [10]) involves, as in Section 5, a careful Morse analysis of the critical points at infinity together with an Euler-Poincaré characteristic argument. In some sense, it suggests that a combination of topological properties of Ω (or M) and topological properties of the function $K(x)$ may force existence. Much work remains to be done in that direction.

6.4. Neumann boundary condition.

Little is known if one replaces in Problem (I) the Dirichlet boundary condition by a Neumann condition, for example,

$$\frac{\partial u}{\partial n} = 0 \quad \text{on} \quad \partial\Omega$$

or by a nonlinear boundary condition, such as

$$-\frac{\partial u}{\partial n} = u^q \quad \text{on} \quad \partial\Omega$$

where $q = (N + 1)/(N - 1)$ is the critical Sobolev exponent on the boundary; see however Cherrier [22].

6.5. The case $N = 2$.

When $N = 2$ the critical Sobolev exponent should be replaced by the Trudinger-Moser embedding theorem (see [41] and [32]). Consider, for example the problem

$$\left.\begin{aligned} -\Delta u &= f(u) && \text{on} \quad \Omega \subset \mathbf{R}^2 \\ u &> 0 && \text{on} \quad \Omega \\ u &= 0 && \text{on} \quad \partial\Omega \end{aligned}\right\} \qquad (17)$$

where f is a positive function on $(0,\infty)$ such that $f(0) = 0$, $0 \leqslant f'(0) < \lambda_1$ and $f(u)$ behaves, as $u \to \infty$, like $e^{\alpha u^2}$ (or $ue^{\alpha u^2}$) ($\alpha > 0$).

Unexpected existence results have been obtained by Carleson-Chang [19] and Atkinson-Peletier [6]. It is not clear, at the moment, where is the border line between existence and nonexistence. It would be interesting to answer the following:

Question 12: Can one find some function f, as above, for which Problem (16) has no solution, say when Ω is a ball?

REFERENCES

[1] A. Ambrosetti and P. Rabinowitz, Dual variational methods in critical point theory and applications, J. Functional Analysis 14 (1973), 349-381.

[2] A. Ambrosetti and P. N. Srikanth, Superlinear elliptic problems and the dual principle in critical point theory, J. Math. Phys. Sci. (to appear).

[3] A. Ambrosetti and M. Struwe, A note on the problem $-\Delta u = \lambda u + u|u|^{2^*-2}$, Manuscripta Math. 54 (1986), 373-379.

[4] F. Atkinson and L. Peletier, Emden-Fowler equations involving critical exponents, J. Nonlinear Anal. (to appear).

[5] F. Atkinson and L. Peletier, Sur les solutions radiales de l'équation, $\Delta u + \frac{1}{2} x \cdot \nabla u + \frac{1}{2} \lambda u + |u|^{p-1}u = 0$, C. R. Acad. Sci. Paris (to appear).

[6] F. Atkinson and L. Peletier, Ground states and Dirichlet problems for $-\Delta u = f(u)$ in R^2 (to appear).

[7] Th. Aubin, Problèmes isopérimétriques et espaces de Sobolev, J. Differential Geometry 11 (1976), 573-598.

[8] Th. Aubin, Equations différentielles nonlinéaires et problème de Yamabe concernant la courbure scalaire, J. Math. Pures Appl. 55 (1976), 269-293.

[9] A. Bahri, Un problème variationnel sans compacité dans la géométrie de contact, C. R. Acad. Sci. Paris 299 (1984), 757-760 and detailed paper to appear.

[10] A. Bahri and J. M. Coron, Une théorie des points critiques à l'infini pour l'équation de Yamabe et le problème de Kazdan-Warner, C. R. Acad. Sci. Paris 300 (1985), 513-516 and detailed paper to appear.

[11] A. Bahri and J. M. Coron, Sur une équation elliptique nonlinéaire avec l'exposant critique de Sobolev, C. R. Acad. Sci. Paris 301 (1985), 345-348 and detailed paper to appear.

[12] J. P. Bourguignon and J. P. Ezin, Scalar curvature functions in a conformal class of metrics and conformal transformations (to appear).

[13] H. Brezis, Periodic solutions of nonlinear vibrating strings and duality principles, Bull. Amer. Math. Soc. 8 (1983), 409-426.

[14] H. Brezis, Elliptic equations with limiting Sobolev exponents - the impact of topology, Proc. Symp. 50th Anniversary of the Courant Institute, Comm. Pure Appl. Math. (to appear).

[15] H. Brezis and J. M. Coron, Convergence of solutions of H-systems or how to blow bubbles, Arch. Rational Mech. Anal. 89 (1985), 21-56.

[16] H. Brezis, J. M. Coron and L. Nirenberg, Free vibrations for a nonlinear wave equation and a theorem of P. Rabinowitz, Comm. Pure Appl. Math. 33 (1980), 667-689.

[17] H. Brezis and L. Nirenberg, Positive solutions of nonlinear elliptic equations involving critical Sobolev exponents, Comm. Pure Appl. Math. 36 (1983), 437-477.

[18] A. Capozzi, D. Fortunato and G. Palmieri, An existence result for nonlinear elliptic problems involving critical Sobolev exponents, Ann. Inst. Poincaré, Analyse Nonlinéaire (to appear).

[19] L. Carleson and S. Y. Chang, On the existence of an extremal function for an inequality of J. Moser (to appear).

[20] G. Cerami, D. Fortunato and M. Struwe, Bifurcation and multiplicity results for nonlinear elliptic problems involving critical Sobolev exponents, Ann. Inst. Poincaré, Analyse Nonlinéaire 1 (1984), 341-350.

[21] G. Cerami, S. Solimini and M. Struwe, Some existence results for superlinear elliptic boundary value problems involving critical exponents, J. Functional Analysis (to appear).

[22] P. Cherrier, Problèmes de Neumann nonlinéaires sur les variétés riemanniennes, J. Functional Analysis 57 (1984), 154-206.

[23] F. Clarke and I. Ekeland, Hamiltonian trajectories having prescribed minimal period, Comm. Pure Appl. Math. 33 (1980), 103-116.

[24] J. Escobar, Positive solutions for some semilinear elliptic equations with critical Sobolev exponents (to appear).

[25] J. Escobar and R. Schoen, Conformal metrics with prescribed scalar curvature (to appear).

[26] M. Escobedo and O. Kavian, Variational problems related to self-similar solutions of the heat equation, J. Nonlinear Anal. (to appear).

[27] D. Fortunato and E. Jannelli, Infinitely many solutions for some nonlinear elliptic problems in symmetrical domains (to appear).

[28] J. Kazdan, Prescribing the Curvature of a Riemannian Manifold, CBMS Regional Conference, Vol. 57, American Mathematical Society (1985).

[29] J. Kazdan and F. Warner, Scalar curvature and conformal deformations of Riemannian structure, J. Differential Geometry 10 (1975), 113-134.

[30] P. L. Lions, The concentration-compactness principle in the calculus of variations, the limit case, Rev. Mat. Iberoamericana 1 (1985), 145-201 and 45-121.

[31] B. McLeod (in preparation).

[32] J. Moser, A sharp form of an inequality by N. Trudinger, Indiana Univ. Math. J. 20 (1971), 1077-1092.

[33] W. M. Ni, On the elliptic equation $\Delta u + K(x)u^{(n+2)/(n-2)} = 0$, its generalizations and applications in geometry, Indiana Univ. Math. J. 31 (1982), 493-529.

[34] S. Pohozaev, Eigenfunctions of the equation $\Delta u + \lambda f(u) = 0$, Soviet Math. Dokl. 6 (1965), 1408-1411.

[35] P. Pucci and J. Serrin, A general variational identity, Indiana Univ. Math. J. (to appear).

[36] J. Sacks and K. Uhlenbeck, The existence of minimal immersions of 2-spheres, Ann. Math. 113 (1981), 1-24.

[37] R. Schoen, Conformal deformation of a Riemannian metric to constant scalar curvature, J. Differential Geometry 20 (1984), 479-495.

[38] S. Solimini, On the existence of infinitely many radial solutions for some elliptic problems (to appear).

[39] M. Struwe, A global compactness result for elliptic boundary value problems involving limiting nonlinearities, Math. Z. 187 (1984), 511-517.

[40] C. Taubes (in preparation).

[41] N. Trudinger, On imbedding into Orlicz spaces and some applications, J. Math. Mech. 17 (1967), 473-484.

[42] D. Zhang, On multiple solutions of $\Delta u + \lambda u + u|u|^{4/(n-2)} = 0$, (to appear).

Département de Mathématiques
Université P. et M. Curie
4 pl. Jussieu
75252 Paris Cedex 05, France

The Differentiability of the Free Boundary for the n-Dimensional Porous Media Equation

Luis A. Caffarelli

In this lecture we want to discuss joint work with N. Wolanski on the question of the differentiability of the free boundary for solutions of the n-dimensional porous media equation

$$\Delta u^m - u_t = 0 \ . \qquad (1)$$

It is well-known that if one starts, at time $t = 0$, with a smooth compactly supported initial datum

$$u(x,0) = \varphi(x)$$

then there exists a unique weak solution $u(x,t)$ of (1) assuming this initial datum and the support $\text{supp}\ u(\cdot,t)$ of $u(\cdot,t)$ is compact for $t > 0$ and increases with t.

One interpretation of this degenerate parabolic equation is that of a gas, with density u, propagating along a porous media according to the laws

$$\vec{v} = -\nabla p$$

and

$$p = u^{\gamma}$$

where $\vec{v}$ is the velocity of p is the pressure.

Conservation of mass then gives us

$$\text{div}(\rho\vec{v}) = \rho_t$$

or

ISBN 0-12-195255-X

$$\Delta u^{\gamma+1} = C(\gamma)u_t$$

which we normalize to

$$\Delta u^m = u_t.$$

This interpretation indicates that if one wants to study the mechanics of the free boundary displacement one should concentrate his attention on $p = u^{m-1}$, since the speed of the free boundary is expected to be $|\nabla p|$. That is, under ideal circumstances the free boundary

$$\partial(\text{supp } p(\cdot,t))$$

should be a smooth surface, p should be smooth up to it, and the free surface should move with speed $|\nabla p|$.

Of course, these ideal circumstances should not necessarily hold from time $t = 0$, since supp φ may be topologically complicated and advancing free boundaries may hit each other.

On the other hand, if supp φ is contained in a convex set - say the unit ball B_1 - it is our belief that as soon as the free boundary lies entirely outside of $B_{1+\varepsilon}$ for some $\varepsilon > 0$ it should be a smooth surface advancing at a strictly positive finite speed.

In two recent papers ([C-V-W] and [C-W]), it is proved that this is the case under the extra transversality hypothesis: $\varphi^{m-1} \in C^1(\text{supp } \varphi)$ and $\nabla\varphi^{m-1} \neq 0$ along $\partial\text{supp } \varphi$.

In fact, in [C-V-W] it is proved that under this extra transversality hypothesis,

a) $p = v = u^{m-1}$ is Lipschitz and non-degenerate up to the free boundary:

$$C_1 d(x,F_t) < v(x,t) < C_2 d(x,F_t)$$

where F_t is the free boundary at time t and d denotes distance.

b) F_t is a locally Lipschitz surface, and in fact $D_\mu p < 0$ on F_t for any direction μ close enough to radial.

c) F_t advances with a finite and strictly positive speed, that is if $y \in F_{t+h}$, then $d(Y,F_t) \sim h$.

In [C-W] the analysis starts from the local situation attained in [C-V-W] and proceeds to show that under these circumstances F_t is really a $C^{1,\alpha}$ surface in space and a C^1 surface in time.

We now discuss the basic ideas involved in the proof of these results. For this purpose we recall that v satisfies an equation of the form

$$v\Delta v + C|\nabla v|^2 = v_t \ ,$$

and that Δv is bounded from below by the well-known Aronson-Benilan estimate ([A-B]), and that the porous medium equation is invariant under the first order scaling

$$\bar{v}(x,t) = \frac{1}{\lambda} v(\lambda x, \lambda t) \ .$$

Therefore, if we fix our attention at a free boundary point (x_0,t_0) (= (0,0)), after an appropriate scaling, etc., we are able to start with the following local geometric configuration.

a) $v(x,t)$ is defined in $B_1(0) \times [-1,0]$, Lipschitz in x and t and $\Delta v > -\varepsilon$ (note that Δv is not invariant under the scaling and we can make its lower bound small by considering a small enough neighborhood of (x_0,t_0)).

b) $D_\mu v > 0$ for $\alpha(\mu,e_n) < \alpha_0$ (where $\alpha(\cdot,\cdot)$ denotes the angle between two vectors and e_n is the unit vector in the nth coordinate direction).

c) $D_{e_n} v$, $D_t v > \delta_0 > 0$ for $v > 0$.

We now assert that in a neighborhood of (0,0) the free boundary is $C^{1,\varepsilon}$ in space and C^1 in time.

To prove it we look at the problem this way: First we think of property b) as saying "for any small translation $\varepsilon\mu$, with μ in the cone

$$\Gamma(\alpha_0,e_n) = \{\mu | \alpha(\mu,e_n) < \alpha_0\}$$

we have

$$v(x + \varepsilon\mu),t) > v(x,t)" \ .$$

Now for some translations μ in the cone $\Gamma(\alpha_0,e_n)$ it may well be that this inequality cannot is not strict, or at least its "strictness" cannot be estimated.

For instance if v is a linear front with a free boundary tangent to the cone $\Gamma(\alpha_0, e_n)$, the inequality will actually be an equality for μ a displacement in the direction of tangency. But if the direction of the displacement is chosen properly, one may expect the inequality to be strict at least away from the free boundary.

How to chose such a direction? For that, we should look at some point away from the free boundary, and somewhat in the past - say $(\frac{1}{4} e_n, -\gamma_0)$ with γ_0 small, and see in which direction the (spatial) gradient of v is pointing.

As in the example of the advancing front, it may well be that the cone $\Gamma(\alpha_0, e_n)$ is tangent to the half space of directions, $H = \{\mu | \langle \mu, \nabla v(\frac{1}{4} e_n, -\gamma_0)\rangle > 0\}$.

But at least we can say that for any unit vector μ, in H

$$D_\mu v(\frac{1}{4} e_n, -\gamma_0) > Cd(\mu, \partial H) .$$

If further, $\mu \in \Gamma(\alpha_0, e_n)$, $D_\mu v$ is everywhere non-negative, by hypothesis and hence, by Harnack's inequality, near $(\frac{1}{4} e_n, -\gamma_0)$, upwards in time we obtain that

$$D_\mu v > Cd(\mu, \partial H)$$

and in particular for $-\frac{3}{4}\gamma_0 < t < -\frac{1}{2}\lambda_0$ for x close to $\frac{1}{4} e_n$,

$$v(x + \varepsilon\mu, t) > v(x,t) + (\varepsilon d(\mu, \partial H) .$$

We now ask, will this inequality propagate to the free boundary? In other words, is it true that given two Lipschitz non-degenerate solutions $v_1 < v_2$, such that $v_1 < v_2 - \varepsilon$ in some region where both are positive, their free boundaries will drift ε-away after some time?

If the answer is yes, we will be able to say that on, say, $B_{1/4} \times [-\gamma_0/4, 0]$ not only

$$v(x + \varepsilon\mu, t) > v(x,t)$$

but further, that

$$v(x + \varepsilon(\mu + \delta\zeta), t) > v(x,t)$$

for any unit vector ζ, provided that δ is a small multiple of $d(\mu, \partial H)$, that is $D_\sigma v > 0$ for $\sigma = \mu + \delta\zeta$ with μ in $\Gamma(a_0, e_n)$, $|\zeta| = 1$ and δ as above.

This enlarged family of directions contains an intermediate cone

$$\Gamma(\alpha_0, e_n) \subset \Gamma(\alpha_1, e^*) \subset H$$

with aperture $\alpha_1 = (1 - s)\alpha_0 + s(\pi/2)$ and $s > 0$ depending only on the constants above.

Rescaling and repeating the argument above inductively, we end up with a monotone family of cones,

$$\Gamma(\alpha_k, e^k)$$

with apertures $\alpha_k > \pi/2 - \lambda^k (\lambda < 1)$ such that

$$D_\mu v > 0$$

in $B_{A^{-k}} \times [-B^{-k}, 0]$ (A and B large enough) and this implies the $C^{1,\alpha}$ space regularity of the free boundary.

Time regularity then follows from a blowup argument that reduces the problem to the one dimensional case.

It only remains to discuss the fact that if one "lifts" the data of a "good" solution v by ε, the free boundary drifts ε-away in the near future (by good we mean Lipschitz and non-degenerate near the free boundary and with a small lower bound for Δv).

This is attained in two steps,

i) If v satisfies these conditions and $v_\varepsilon = v + \varepsilon(v - af(x) + Bt)^+$ with ε, a, b small, $f(x) \equiv 0$ on $B_{1/2}$ and $f(x) \equiv 1$ outside B_1 then v_ε is a subsolution on

$$\{B_1 \times [-1, 0]\} \cap \{v < \gamma\} \quad (\gamma \text{ small}) .$$

The reason being that for $L(w) = w\Delta w + C(\nabla w)^2 - w_t$

$$L((1 + \varepsilon)v) = (1 + \varepsilon)L(v) + \varepsilon(v\Delta v + (\nabla v)^2) + O(\varepsilon^2) > C\varepsilon$$

(since Δv is bounded from below) allowing for the correction terms f and t, and

ii)

$$L((1 + \varepsilon)v(x + C\varepsilon(t + 1)e_n, t) = 1 + \varepsilon)L(v) + \varepsilon(v\Delta v|\nabla v|^2)$$
$$- C\varepsilon D_n v > B\varepsilon$$

if C is chosen small allowing again for a correction in x.

REFERENCES

[A-B] D. Aronson and Ph. Benilan, Régularité des solutions de l'équation des milieux poreux dans R^n, C. R. Acad. Sci. Paris Sér. A-B 288 (1979), 103 - 105.

[C-V-W] L. A. Caffarelli, J. L. Vazquez and N. Wolanski, Lipschitz continuity of solutions and interfaces of the n-dimensional porous medium equation IMA preprint #191, (1985).

[C-W] L. A. Caffarelli and N. Wolanski, The differentiability of the free boundary for the n-dimensional porous media equation.

Department of Mathematics
University of Chicago
Chicago, IL 60637

Oscillations and Concentrations in Solutions to the Equations of Mechanics

Ronald J. DiPerna

We are concerned with the analysis of the static structure and dynamic behavior of oscillations and concentrations in compressible and incompressible media. In the context of compressible fluids and solids the topics involve the qualitative behavior of solutions to hyperbolic systems of conservation laws in one space-dimension,

$$\partial_t u + \partial_x f(u) = 0 \ . \tag{1}$$

In the context of incompressible fluids the topics involve solutions to the Euler equations in two space dimensions,

$$\begin{aligned} \partial_t u + u \cdot \nabla u + \nabla p &= 0 \\ \operatorname{div} u &= 0 \ . \end{aligned} \tag{2}$$

The goal for systems (1) and (2) is to analyze oscillations and concentrations in solutions to the initial value problem, specifically to analyze weak limits of solutions.

The main focus is on two classes of problems. In the first, the system is fixed and the data sequence contains oscillations. Here we are concerned with the sequential analysis of solutions corresponding to sequences of initial data $u_0^\varepsilon(x)$ which contain oscillations and concentrations. The parameter ε may stand for any of a variety of quantities, e.g. a ratio of length scales, an error tolerance in the data, etc. The objective is to describe the dynamic response in terms of the formation, modulation, interaction and decay of

ISBN 0-12-195255-X

oscillations and concentrations. What are the mechanisms which reduce, enhance or modulate oscillations in the solution as time evolves? Are oscillations in the data annihilated in a small initial layer? If they survive, how do they interact in the short and long run?

The second class of problems deals with systems of equations which change with ε while the initial data is held fixed. In this situation ε typically represents a diffusion or dispersion parameter in the form of a small coefficient multiplying a higher order term. What are the mechanisms which reduce, enhance or modulate oscillations in the zero diffusion limit and zero dispersion limit? We are concerned here mainly with the analysis of solutions to the initial value problem for singularly perturbed hyperbolic systems of conservation laws in one space dimension of the form

$$\partial_t u + \partial_x f(u) = \varepsilon \partial_x^2 u \quad , \tag{3}$$

and with the Navier-Stokes equations in two space dimensions:

$$\begin{aligned} \partial_t u + u \cdot \nabla u + \nabla p &= \varepsilon \Delta u \\ \operatorname{div} u &= 0 \\ u = u^{\varepsilon}(x,t), \quad u^{\varepsilon}(x,0) &= u_0(x) \ . \end{aligned} \tag{4}$$

For both of the processes (3) and (4) the notion of measure-valued solution to the underlying equations (1) and (2) is relevant to the representation and analysis of weakly convergent solution sequences. We shall briefly recall the definition of measure-valued solution and then discuss some results which have been obtained dealing with convergence of singular perturbations. For simplicity we shall state the definition in the context of systems of conservation laws in one space dimension. The formulation is similar in several space dimensions for both the compressible and incompressible equations.

Definition. A mapping

$$(x,t) \longrightarrow \nu_{(x,t)}$$

from the physical domain R^2 to the space of probability measures $\mathrm{Prob}(R^n)$ over the statespace R^n is called a measure-valued solution for a system

$$\partial_t u + \partial_x f(u) = 0, \quad u \in R^n \tag{5}$$

of n conservation laws if the time derivative of the center of mass,

$$\langle \nu_{(x,t)}, \lambda \rangle \equiv \int_{R^n} \lambda d\nu_{(x,t)} ,$$

is balanced by the spatial derivative of the expected value of the flux f,

$$\langle \nu_{(x,t)}, f(\lambda) \rangle \equiv \int_{R^n} f(\lambda) d\nu_{(x,t)} ,$$

i.e., if

$$\partial_t \langle \nu_{(x,t)}, \lambda \rangle + \partial_x \langle \nu_{(x,t)}, f(\lambda) \rangle = 0 .$$

This definition can be contrasted to the classical definition of solution to (5) in the following fashion. A function

$$(x,t) \rightarrow u(x,t) \in R^n$$

from the physical space to the state space R^n is a solution of (5) if the divergence of the vector-field $\{u(x,t), f(u(x,t))\}$ vanishes. In this classical situation, one associates with each point of the physical domain an n-tuple of numbers. More generally, one may associate with each point of the physical domain a measure over the state space. Such a map qualifies as a measure-valued solution of the divergence of the expected value if the field (u,f(u)) vanishes.

The link between classical solutions and measure-valued solutions is the following. Classical solutions correspond to measure-valued solutions which take the form of Dirac masses: a mapping

$$\nu: (x,t) \rightarrow \delta_{u(x,t)}$$

is a measure-valued solution of (5) if and only if the point of concentration u(x,t) is a classical solution. Of course, when speaking of solutions in either the classical or generalized sense, we interpret derivatives in the sense of distributions. Both equations (5) and (6) are understood to hold in the sense of distributions.

The main unifying feature of the notion of measure-valued solution in the context of general limiting processes is the following. If $\{u^\varepsilon\}$ is any sequence of solutions to a perturbed pde whose amplitude is uniformly bounded with respect to the perturbation parameter ε in the sense that

$$\int |u^{\varepsilon}(x,t)|^{p} dx \leq \text{constant} ,$$

then there exists a measure-valued solution of the underlying equation which describes the weak limit of all state variables. If g is an arbitrary continuous function on the state space then

$$\text{weak-}\lim_{\varepsilon\to 0} g(u^{\varepsilon}(x,t)) = \int_{R^n} g(\lambda)\, d\nu_{(x,t)} .$$

The weak limit of the composition coincides with the expected value of g. Here the measure-valued solution is given by the associated Young measure.

In brief, if both the perturbed and unperturbed systems have divergence form and if the problem is L^p-stable, then a measure-valued solution ν is generated to the unperturbed equation. The general goal is to describe the structure of measure-valued solutions associated with various limiting processes, e.g. diffusion processes and dispersion processes.

The first step deals with the circumstances under which no substantial oscillations develop as the parameter ε tends to zero, in the sense that solution sequence u^{ε} under investigation converges in norm. It turns out that strong convergence corresponds to the situation where the measure-valued solution of the underlying equation reduces to a Dirac mass at each point (x,t). Strong convergence means minimal support for the associated measure-valued solutions. In general, the deviation between weak and strong convergence is reflected in the spreading of the support of the measure-valued solution. The type of oscillations which arise in the limit as the perturbation parameter varies is encoded in the structure of the measure-valued solution.

We shall next recall several results and conjectures dealing with measure-valued solutions. Let us first consider the setting provided by a hyperbolic system of conservation laws in one space dimension as typified by the equations of 1-dimensional compressible gas dynamics. In this situation one anticipates that the measure-valued solutions associated with diffusive processes reduce to Dirac masses, i.e. the convergence is strong.

One of the main conjectures is that solutions of the compressible Navier-Stokes equations converge to the solution of the compressible Euler equations as the viscosity coefficient vanishes, equivalently as the Reynolds number approaches infinity. No substantial oscillations should develop to preclude close agreement of inviscid and viscous equations. In the setting of the 1-dimensional equations, this conjecture has been verified [2,3] with the aid of the recently developed theory of compensated compactness initiated by the work of L. Tartar and F. Murat [7,9,10]. The reason for the strong convergence for this type of diffusive process stems from the entropy condition: the second law of thermodynamics is responsible for suppressing substantial oscillations in a compressible gas.

The aforementioned result generalizes to a broad class of systems of two nondegenerate conservation laws in one space dimension. Perturbed solution sequences generated by parabolic regularizing operations converge strongly to solutions of the unperturbed hyperbolic system. There is interest in extending these results to general systems in one space dimension and to specific physical systems in several space dimensions.

With regard to historical background, it is appropriate to remark that the development of the functional analytic theory of conservation laws began in the mid-seventies with the basic work of L. Tartar on a scalar conservation law in one space dimension [10]. Further study of oscillations in solutions to systems of conservation laws has led to the rigorous results dealing with the convergence of classical first-order finite difference schemes and diffusive processes based on parabolic operators as in compressible Navier-Stokes [2,3].

In contrast to diffusion processes, dispersion processes for hyperbolic equations generate measure-valued solutions which do not have Dirac structure. For example, the measure-valued solution of the inviscid Burgers equation,

$$\partial_t u + \partial_x (u^2/2) = 0 \ , \tag{7}$$

which represents the zero dispersion limit for the KdV equation flows as a Dirac mass up to the time that shock waves form at the hyperbolic level. The convergence of KdV solutions is strong up to the time when acoustic waves in equation (7) focus to produce a shock. Thereafter, the measure-valued solution spreads in accordance with the formation of oscillations in KdV. The work of P. Lax, D. Levermore and S. Venakides [5,11] has provided descriptions of the motion of the first several moments of the associated measure-valued solution together with detailed information dealing with the creation and annihilation of phases.

Effort has been directed to several situations where initial oscillations and concentrations are not completely suppressed by the dynamics. This situation is typical of incompressible fluids. Next we shall briefly describe a program with A. Majda concerning incompressible flows with highly concentrated vorticity. Consider the 2-dimensional Euler equations (2) with initial data having finite kinetic energy and finite total vorticity. Specifically, we are concerned with the case where the initial vorticity is represented by a singular measure as in the case of flows with vortex sheets.

There are several questions and problems: Does there exist a globally defined distributional solution to the (conservative) equations,

$$\begin{aligned} \partial_t u + \operatorname{div} u \otimes u + \nabla p &= 0 \\ \operatorname{div} u &= 0 \ , \end{aligned}$$

if the initial data have finite kinetic energy and finite total vorticity? In this setting considerations of temporal regularity are paramount. Does there exist a global solution whose energy changes continuously with time? Should other levels of temporal regularity be considered in order to entertain the possibility of inviscid energy dissipation?

A related problem for the Navier-Stokes equations is to determine whether or not solutions u^ε of the 2-dimensional system (4) converge in the energy norm, i.e. in L^2 or L^2_{loc}, to a solution of the 2-dimensional Euler equations as the Reynolds number R approaches infinity, i.e. as ε approaches zero.

In this connection it is appropriate to remark that if the initial vorticity lies in L^p with $p > 1$, then no functional analytic difficulties arise in proving existence of global

Euler solutions and strong L^2 convergence of Navier-Stokes in two space dimensions. The operation of reconstructing the stream function ψ from the vorticity ω at a fixed time t through the potential equation gains two derivatives. At each fixed time the velocity field lies in a compact set of L^2. Temporal variations are not difficult to handle.

Here we are interested in the problem of representing and analyzing oscillations and concentrations in single Euler solutions and sequences of Euler solutions when the vorticity lies in the space of bounded measures. Within this setting two uniform estimates are available. Uniform control on the kinetic energy and total vorticity. If

$$\partial_t u^\varepsilon + \operatorname{div} u^\varepsilon \otimes u^\varepsilon + \nabla p^\varepsilon = 0$$
$$\operatorname{div} u^\varepsilon = 0$$

then we formally have

$$\int_{R^2} |u^\varepsilon(x,t)|^2 dx = \int_{R^2} |u_0^\varepsilon|^2 dx$$

and

$$\text{total mass } \omega^\varepsilon(\cdot,t) \le \text{total mass } \omega_0^\varepsilon ,$$

where $\omega = \operatorname{curl} u$. Analogous inequalities hold independently of the Reynolds number for 2-dimensional Navier-Stokes.

The interest in concentration questions arises from the fact that, although vorticity is passively transported along particle paths in 2-dimensional flow, the geometry of particle paths may serve to focus vorticity and kinetic energy on small sets. The physical problem is to determine the extent to which concentration effects are compatible with 2-dimensional symmetry. The general program is concerned with both oscillations and concentrations in 2-dimensional flow. We shall limit our remarks here mainly to questions dealing with concentrations. How should they be represented analytically? Do significant concentrations of kinetic energy and vorticity develop in finite time in 2-dimensional Euler flow?

In order to give a somewhat more precise statement of the problem let us consider the Cauchy problem for (2) with fixed initial data having finite kinetic energy and finite total vorticity. Inspecting the solution u at a sequence of times t_n reveals a sequence of velocity fields $u_n = u(\cdot,t_n)$ with uniformly bounded kinetic energy and uniformly bounded total vorticity. The density of kinetic energy serves as the density for a measure K_n whose total mass is uniformly bounded;

$$K_n(E) = \int_E |u_n|^2 dx .$$

Without loss of generality we may assume that K_n converges weakly to a measure K and that the velocity fields u_n converge weakly in L^2 to a field u. The general problem is to analyze concentrations in kinetic energy, i.e. the (possible) loss of compactness in L^2 of the sequence u_n. Is K absolutely continuous with respect to 2-dimensional Lebesgue measure? Can a finite amount of kinetic energy be deposited on a set of zero Lebesgue measure by a 2-dimensional incompressible flow in finite or infinite time? One of the goals is to determine whether or not u_n converges strongly to u in L^2.

More generally, the objective is to analyze limits of general state variables associated with the sequence u_n. Some variables may exhibit a loss of compactness while others avoid it through a subtle process of concentration compensation.

The concentration problem associated with the Cauchy problem for the 2-dimensional Euler equations with fixed initial data having finite kinetic energy and finite total vorticity has several facets. The initial stage of our program deals with the related problem of analyzing concentrations in sequences of 2-dimensional Euler solutions subject to the constraint that the initial data have uniformly bounded kinetic energy and uniformly bounded total vorticity. The strategy here is to introduce a measure-valued solution to the 2-dimensional Euler equations which represents the weak limit of a given sequence of 2-dimensional Euler solutions and then determine the size of the set on which this measure-valued solution reduces to a Dirac mass. The reduction of the measure-valued solution to a Dirac mass on a large set implies that L^2 defects in the velocity field occur only on a small set.

The results so far with A. Majda include several classes of explicit Euler sequences for which the associated measure-valued solution has non-Dirac structure. The explicit sequences represent the collapse of an isolated vortex and are not compact in L^2. They concentrate a finite amount of kinetic energy at a point. To measure the extent of loss of L^2 compactness for general sequences is one of the goals.

The first step in the general goal is to analyze L^2 defects in arbitrary elliptic sequences (ψ_n, ω_n):

$$\Delta\psi_n = \omega_n$$

$$|\psi_n|_{H^1} \leq \text{constant}$$

$$\text{total mass } \omega_n \leq \text{constant} .$$

In the context of fluids, the orthogonal gradient of the stream function ψ generates the velocity field u and lies in L^2. This is the reason for the H^1 hypothesis. The objective is to prove that ψ_n converges strongly in $H^1(E)$, after passing to a subsequence, where E^c is a small set. Such pairs of distributions arise from Euler sequences at a fixed time or from a single Euler solution at a sequence of times.

The study of the elliptic problem above is motivated by questions in the physics of fluids and by the concentration-compactness program of P. L. Lions [12,13]. The program has been concerned with systematic assessment of the losses of compactness in variational problems due to the action of non-compact groups both in the locally compact case and in the limit case. We shall conclude by recalling one motivating example from [12,13] dealing with the loss of compactness related to Sobolev imbedding. Consider a sequence of maps ψ_n which are uniformly bounded in $W^{1,p}$ and hence in L^q where $q = \frac{np}{n-p}$. Without loss of generality we may assume that ψ_n converges weakly to ψ in L^q. The assertion is that ψ_n converges strongly to ψ on the complement of a countable number of points, i.e. the L^q defect is at most a countable set. Similar types of critical structure arise in fluid dynamics and are currently under investigation.

In the context of geometry it is known that concentrations may also be limited for numerous reasons. We refer the reader to the articles of H. Brezis and C. Taubes in this volume and to the references cited therein.

REFERENCES

1. DiPerna, R. J., Convergence of approximate solutions to conservation laws, Arch. Rational Mech. Anal. 82 (1983), 27-70.
2. DiPerna, R. J., Convergence of the viscosity method for isentropic gas dynamics, Comm. in Math. Phys. 91 (1983), 1-30.
3. DiPerna, R. J., Measure-valued solutions to conservation laws, Arch. Rational Mech. Anal. 88 (1985), 223-270.
4. Flaschka, H., Forest, M. G. and McLaughlin, D. W., Multiphase averaging and the inverse spectral solution of the Korteweg-deVries equation, Comm. Pure Appl. Math. 33 (1980), 739-784.
5. Lax, P. D. and Levermore, C. D., The small dispersion limit for the KdV equation. I and II, Comm. Pure Appl. Math. 36 (1983), 253-290, 571-594.
6. McLaughlin, D. W., Modulations of KdV wavetrains, Physica 3D 1 (1981), 355-363.
7. Murat, F., Compacité par compensation, Ann. Scuola Norm. Sup. Pisa 5 (1978), 489-507.
8. Slemrod, M. and Rotyburd, V., Measure-valued solutions to a problem in dynamic phase transitions, to appear.
9. Tartar, L., Compensated compactness and applications to partial differential equations, in Research Notes in Mathematics: Nonlinear Analysis and Mechanics, Heriot-Watt Symposium, Vol. 4 (R. J. Knops, ed.), Pitman Press, 1979.
10. Tartar, L., The compensated compactness method applied to systems of conservation laws, in Systems of Nonlinear Partial Differential Equations (J. M. Ball, ed.), Reidel Publishing Co., 1983.
11. Venakides, S., The zero dispersion limit of the Korteweg-deVries equation with non-trivial reflection coefficient, Comm. Pure Appl. Math. (1984).
12. Lions, P. L., The concentration-compactness principle in the calculus of variations: the locally compact case, Ann. I. H. P. Anal. Non. Lin., 1984.

13. Lions, P. L., The concentration-compactness principle in the calculus of variations: the limit case, Rev. Mat. Iberoamer, 1984.

This research was supported in part by National Science Foundation Grant DMS-8301135.

Department of Mathematics
University of California
Berkeley, CA 94720

The Connection Between the Navier-Stokes Equations, Dynamical Systems, and Turbulence Theory

Ciprian Foias and Roger Témam

INTRODUCTION

A basic tenet of theoretical physics is that Nature is the complicated display of the repetitive interplay of some (mathematically) simple laws. Much human energy, ingenuity and glamour was devoted to the search and the explication of such laws, usually describing physical reality off the reach of our direct experience. Satisfactory theories for these "simple" and "extreme" phenomena seem to have been attained several times in this century. However, numerous physical phenomena occuring at the scale of our daily experience and hourly impinging on our lives remain still beyond our present theoretical understanding. Among these phenomena the most conspicuous are those displaying turbulence.

What is turbulence? Loosely speaking, turbulence is an irreversible disorder, statistically "organized" according to several time and space scales, of mostly fast moving fluid flows. There are three major distinct but not mutually contradictory theoretical views on turbulence, namely:

ISBN 0-12-195255-X

a) The statistical view, in which the turbulence is considered as the observed behaviour of the evolution of statistical distributions of flows instead of the evolution of one individual flow.

b) The viewpoint of regularity breakdown, in which the turbulence is considered to result from the blow-up of the vorticity in finite time, albeit necessarily on a set of small Hausdorff dimension. This might express the fact that the continuum mechanics model could become invalid at least intermittently in time.

c) The dynamical systems view, in which the turbulence is a phenomenologic perception of the long time complicated behaviour of the individual flows.

Among the most influential proponents of these views were: O. Reynolds [1], G. I. Taylor [2], T. von Kármán [3], and A. N. Kolmogorov [4] for a); J. Leray [5] and B. Mandelbrot [6] for b); and L. L. Landau [7], E. Hopf [8] and D. Ruelle-R. Takens [9] for c). Although a) seems to be the view of most aeronautical engineers, mathematicians and theoretical physicists seem to hold the other views (especially c)). There are two reasons for this fact. First, b) and c) have inspired interesting mathematical research ([10], [11], [12], [13], [14], [15], [16],...) and some illuminating mathematical metaphors for the onset of turbulence under very stringent physical conditions ([17], [18], [19], [20], [21],...). Notice however that the phenomenon of turbulence cannot be equated with that of the onset of turbulence.

It is premature to exclusively adopt one of the views a), b), or c), but it is reasonable to test them by trying to establish rigorous mathematical facts based on the Navier-Stokes equations, the interpretation of which are consistent with any of a), b), and c). In this paper we shall present our contribution to this program from the dynamical systems point of view c). Moreover we prove in Section 3 a regularity and backward uniqueness theorem on an open dense subset of the universal attractor of the 3D Navier-Stokes equations. As far as we know this theorem is new and is the first of its kind.

In dedicating this paper to the sixtieth anniversary of Professor John Nohel, the authors express to him their warm and high appreciation.

2. THE NAVIER-STOKES EQUATIONS AND THEIR ABSTRACT FRAMEWORK

In this Section we shall recall some basic facts on the Navier-Stokes equations which concern a larger class of related dissipative partial differential equations (see for instance [22], [23]). First we consider the Navier-Stokes equations.

$$\left.\begin{aligned} &u_t + (u\cdot\nabla)u = \nu\Delta u - \nabla p + f \\ &\nabla\cdot u = 0 \end{aligned}\right\} \quad \text{in } \Omega\times(0,\infty) \subset \mathbb{R}^{n+1} \tag{2.1}$$

where Ω is an open connected bounded set $\subset \mathbb{R}^n$ $(n = 2 \text{ or } 3)$ such that its boundary Γ is a compact manifold of class $\mathcal{C}^2$ of dimension $n - 1$, $\Gamma = \partial\bar{\Omega}$ or $\Omega = \prod_{j=1}^{n} [a_j, b_j]$. In the first case (2.1) is supplemented with homogeneous boundary conditions

$$u|_\Gamma = 0 \tag{2.2}$$

while in the second case it is supplemented with the periodic and zero mean conditions

$$u|_{x_j=a_j} = u|_{x_j=b_j} \quad (1 \le j \le n) \quad , \quad \int_\Omega u = 0 \quad . \tag{2.2}$$

In both cases let V be the space of the divergence free $u \in H^1(\Omega)^n$ satisfying the respective boundary conditions, let H be the closure of V in $L^2(\Omega)^n$ and let P denote the orthogonal projection of $L^2(\Omega)^n$ on H . Introducing the operators

$$Au = -P\Delta u \ , \ B(u,v) = P[(u\cdot\nabla)v] \ , \ u,v \in V\cap H^2(\Omega)^n \ , \tag{2.3}$$

the equations (2.1)-(2.2), (2.2)$'$ become a particular case of the abstract dissipative evolution equations of the type

$$\frac{du}{dt} + \nu Au + R(u) = 0 \tag{2.4}$$

$$R(u) = B(u,u) + C(u) - f \ , \tag{2.5}$$

where $A : \mathfrak{D}_A \to H$ is a self-adjoint operator, with compact inverse, in a Hilbert space H , B and C are bilinear, resp. linear, operators from $\mathfrak{D}_A$ into H , $\nu > 0$, and $f \in H$ are fixed. The operators B and C satisfy usually conditions of the form

$$\begin{aligned} |B(u,v)| &\le c_1|A^\alpha u||A^\beta v| \qquad u,v \in \mathfrak{D}_A \\ |C(u)| &\le c_2|A^\gamma u| \qquad u \in \mathfrak{D}_A \end{aligned} \tag{2.6}$$

where $|\cdot|$ denotes the norm in H , α , β , $\gamma \in [0,1)$ and $\alpha + \beta$ is usually fixed. For instance, for the equations (2.1)-(2.2), (2.2)$'$, $\alpha + \beta = 1/2 + n/4$ and $C = 0$. (Here and in the sequel $c_1, c_2, \ldots$ will denote adequate absolute constants.)

Many other equations more or less related to the Navier-Stokes equations can be easily given under appropriate boundary conditions the form (2.4). We mention the Boussinesq equations, the Kuramoto-Sivashinsky equations, the (local or nonlocal) Burgers equations, etc. We only notice that the special form (2.5) of $R(u)$ is not a necessary feature of (2.4). Essentially we need that

$$\left|\frac{\partial R(u)}{\partial u} v\right| \leq c_1' \; |A^{\alpha}u||A^{\beta}v| \qquad (u,v \in \mathfrak{D}_A) \quad , \tag{2.6$'$}$$

where $\partial R(u)/\partial u$ is a Gateaux type derivative and α , β are as in (2.6). In this way the Kahn-Hilliard equations and some diffusion-reaction equations can also be given a similar although a bit more complicated form. However, we shall consider here only the case (2.4), (2.5), (2.6), supplemented with the condition

$$(B(u,v),v) = 0 \quad , \quad u,v \in \mathfrak{D}_A \tag{2.7}$$

which is verified by (2.1)-(2.2), (2.2)$'$; moreover in the case $n = 2$ and (2.1)-(2.2)$'$, $B(u,u)$ also verifies

$$(B(u,u),Au) = 0 \qquad u \in \mathfrak{D}_A \quad . \tag{2.7$'$}$$

(The relations (2.7) and (2.7)$'$ reflect the energy, resp. enstrophy, conserving property of the inertial forces in (2.1).)

Finally let us recall that the initial value problem for (2.4) has always a (weak) solution on $[0,\infty)$. By a (weak)

solution on a time interval σ we mean a function
$u(\cdot) \in C(\sigma; H_{weak})$ satisfying in V' the integral form of
(2.4), namely

$$u(t) = u(t_0) + \int_{t_0}^{t} \left[\nu Au(\tau) + R(u(\tau))\right] d\tau \quad (t_0 < t, \ t_0, \ t \in \sigma),$$

and satisfying the energy inequality

$$\frac{1}{2} |u(t')|^2 + \nu \int_{t}^{t'} \|u(t)\|^2 dt \leq \frac{1}{2} |u(t)|^2 + \int_{t}^{t'} (f, u(t)) \ dt \quad (2.8)$$

for all $t' \in \sigma$, $t' \geq t$ and t a. e. in σ, and such that

$$u \in L^{\infty}(\sigma; H) \cap L^2_{loc}(\sigma; V) \quad . \tag{2.9}$$

For every weak solution in σ there is an open set σ_u
of total measure in σ such that u is regular on the intervals of σ_u . The (weak) solution is called regular on σ
if moreover $u \in C(\sigma; V)$. This solution is then an analytic
$\mathfrak{D}_A$ -valued function (see for instance [14]).

In all the mentioned cases except those involving the 3D
Navier-Stokes, u is uniquely determined by u_0 and $\sigma_u =$
$(0, \infty)$. In this case as well as in the case when the value
$u(t)$ is uniquely determined by u_0 we shall also denote it
by $S(t, u_0)$. (Thus $S(0, u_0) = u_0$.) The regularity
break-down viewpoint b) is based on the assumption that for
the 3D Navier-Stokes equations there exist weak solutions on
$(0, \infty)$ which are not regular.

3. THE UNIVERSAL ATTRACTOR

Let X denote the set of those $u_o \in H$ for which there exists a weak solution on $\sigma = (-\infty,\infty)$ of (2.4), H-bounded on $(-\infty,\infty)$, such that $u(0) = u_o$; $X \neq \phi$, since (2.4) has stationary solutions. From (2.7), (2.9) it follows readily

$$X \subset \left\{u \in H : |u| \leq \frac{|A^{-1/2}f|}{\nu\lambda_1^{1/2}}\right\} . \tag{3.1}$$

The definition obviously implies that if $u_o \in X$ there exists at least one solution $u(t)$ such that $u(0) = u_o$ and $u(t) \in X$ for all t ; thus in particular

$$S(t,u) \in X \quad \text{whenever} \quad u \in X \quad , \quad -\infty < t < \infty \tag{3.2}$$

and $S(t,u)$ makes sense. Moreover for every weak solution $u = u(\cdot)$ of (2.4) in $(0 , \infty)$, we have

$$u(t) \to X \quad \text{in} \quad H_{weak} \quad \text{for} \quad t \to \infty \tag{3.3}$$

(i.e. every weak neighbourhood of X will eventually contain $u(t)$). This can be shown by a straightforward argument based on the following

<u>Lemma.</u> Let $\{u_j(\cdot)\}$ be a sequence of weak solutions on $\sigma = (a,b)$ bounded in H Then there exists a subsequence $\{u_{j'}(\cdot)\}$ and a weak solution $u(\cdot)$ on σ such that $u_{j'}(t) \to u(t)$ in H_{weak} uniformly on any compact sub-

interval $\subset \mathcal{O}$. Moreover if $u(\cdot)$ is regular on $\mathcal{O}_1 = (a_1,b_1)$ $(\subset \mathcal{O})$ then $u_{j'}(t) \to u(t)$ in V , uniformly on any compact interval $\subset \mathcal{O}_1$.

For a proof we refer to [24], Ch. 1.

This Lemma also yields easily

$$X \text{ is compact in } H_{\text{weak}} . \tag{3.4}$$

Because of all these simple properties (3.1-4) we call X the <u>universal attractor</u> of the equation (2.4).

A first basic purpose of the dynamical systems approach should be the understanding of the action of $S(\cdot,\cdot)$ on $(-\infty,\infty) \times X$ in this general framework.

In the 2D case (i.e. $n = 2$), the universal attractor satisfies

$$X \text{ is bounded in } V \text{ and } S(t,X) = X \ , \ \forall t . \tag{3.5}$$

Moreover, whenever (3.5) holds, X is included in $\mathcal{D}_A$ and is of finite dimension, i.e. homeomorphic to a compact subset of some $\mathbb{R}^N$ [13], [25], [24]. These properties are necessary if the dynamical systems approach c) is mathematically consistent with the Navier-Stokes equations. Therefore the following open question is basic for both approaches b) and c).

<u>Question 1.</u> Is the universal attractor X included in V even in the 3D case (i.e. $n = 3$)?

It is noteworthy that

$$X \subset V \iff (3.5) . \tag{3.6}$$

The only nontrivial part of (3.6) is

$$X \subset V \implies X \text{ is bounded in } V \ . \qquad (3.7)$$

Proof. If $X \subset V$ but is unbounded in V then there exists $\{u_j\}_{j=0}^{\infty} \subset X$, $u_j \to u_o$ in H_{weak} and $\|u_j\| \to \infty$. By (3.6), there exists $\{v_j\}_{j=0}^{\infty} \subset X$ such that $S(1,v_j) = u_j$, $\forall j \geq 0$. Using the Lemma above, we can also assume that $S(t,v_j) \to u(t)$ in H_{weak} for all $t \in (0,2)$. In particular $u(1) = u_o$ and since $u(t) \in X \subset V$ for all $t \in (0,2)$ we have also, by using once again the Lemma, that $u_j = S(1,v_j) \to u_o$ in V, contradicting the choice of $\{u_j\}_{j=0}^{\infty}$. Thus X must be bounded in V.

It is easy to see that $X \cap V$ is always dense in X (with the topology of H_{weak}). This is not surprising since for any ball $K \subset H$, $K \cap V$ is dense and also meager in K. Therefore the next result is a bit surprising and yields an almost positive answer to Question 1.

Theorem 1. $X \cap V$ contains an open dense subset X_{reg} of X (with the topology of H_{weak}). Moreover for any $u_o \in X_{reg}$ there exists an interval $\sigma_{u_o} = (-a,a)$ such that for any weak solution $u(\cdot)$ satisfying $u(t) \in X$, $\forall t$, we have that $u(\cdot)|\sigma_{u_o}$ is uniquely determined and regular.

Sketch of the Proof. Let X_{reg} be the set of the $u_o \in X$ satisfying the properties given in the statement. We must show that X_{reg} is open and dense in X endowed with the weak topology of H. Using the time analyticity of the regular solutions it is easy to show that if $u_o^o \notin X_{reg}$ then there exists $\{u_j^o\}_{j=1}^{\infty} \subset X$ such that $u_j^o \to u_o^o$ in H_{weak} and

$\|u_j^o\| \to \infty$ (for $j \to \infty$). Obviously this property is also valid for u_o^o in the closure of $X \setminus X_{reg}$. If moreover $u_o^o \in X_{reg}$, and $\{u_j^o\}_{j=1}^{\infty}$ is as above we can, by virtue of the above Lemma, assume that there exist X -valued weak solutions $u_j(\cdot)$ on $(-\infty,\infty)$, $u_j(0) = u_j^o$, $j = 0,1,\ldots$ such that $u_j(t) \to u_o(t)$ in H_{weak} for all $t \in (-\infty,\infty)$. But on the interval $(-a^o,a^o)$ associated to u_o^o described in the statement, $u_o(\cdot)$ is regular. Therefore the Lemma implies that $u_j^o = u_j(0) \to u_o(0) = u_o^o$ in V , contradicting $\|u_j^o\| \to \infty$. Thus $u_o^o \notin X_{reg}$ and $X \setminus X_{reg}$ is closed in X ; hence X_{reg} is open in X .

The density of X_{reg} in X follows easily from the following fact:

Let $u(\cdot)$ be a weak solution on $(-\infty,\infty)$, $u(t) \in X$, $\forall t$, regular on (a,b) . Then $u(c) \in \overline{X}_{reg}$ for any $c \in (a,b)$.

Assuming that $u(c) \notin \overline{X}_{reg}$ for some $c \in (a,b)$, we notice first that there exist $d \in (a,c)$ such that for any X -valued weak solution $v(\cdot)$ on $(-\infty,\infty)$, $v(c) = u(c)$, we have $v(t) \notin \overline{X}_{reg}$, $\forall t \in (d,c]$. (This is a consequence of the equicontinuity of all weak solutions on $(-\infty,\infty)$ with values in X , when viewed as V' -valued.) Therefore if $c_o = c > c_1 > c_2 > \ldots > c_k > c_{k+1} = d$ there must exist a weak solution $v(\cdot)$ on $(-\infty,\infty)$ such that $v(c) = u(c)$ and which is not regular on any interval (c_{j+1},c_j) . It follows that the length of any interval of regularity of $v(\cdot)$ in (d,c) is $\leq 2\delta$ where $\delta = \max_{0\leq j\leq k} (c_j - c_{j+1})$. It is well-known that if $\ell_1,\ell_2,\ldots$ are the length of these intervals then

$\Sigma\sqrt{\ell_j} \leq C$ where the constant C is independent of the X-valued weak solution (see for instance [25]). It follows

$$C\sqrt{2\delta} \geq \Sigma \ell_j \geq c - d - 2\delta$$

Letting $\delta \to 0$ we obtain a contradiction.

This finishes the proof.

4. ESTIMATES OF THE FRACTAL DIMENSION

Let Y be any compact subset of H. The fractal dimension (capacity) $d_M(Y)$ of Y is defined by

$$d_M(Y) = \overline{\lim_{\varepsilon \to 0}} \frac{\log N_\varepsilon(Y)}{\log\frac{1}{\varepsilon}}$$

where $N_\varepsilon(Y)$ is the smallest number of balls of radii $\leq \varepsilon$ covering Y. This dimension, much emphasized by B. Mandelbrot [6] is larger than the classic Hausdorff dimension $d_H(Y)$. It can be ∞ even if the latter one is 0 [26]. Therefore $d_M(Y) < \infty$ is a finer condition of finite dimensionality than $d_H(Y) < \infty$. Moreover if X is the universal attractor of (2.4), $d_M(X)$ can be viewed as a lower bound for the number N of asymptotic degrees of freedom of the phenomena described by the equation (2.4). Indeed the number of real parameters necessary to describe for $t \to \infty$ the asymptotic behaviour of (2.4) must obviously exceed $d_M(X)$. (The reason why the number $N_H \sim 2d_H(X) + 1$ of parameters provided by R. Mañe's theorem (see [16]) may be insufficient is due to the fact that the homeomorphism from a compact set K in $\mathbb{R}^{N_H}$

to X may not be a Hölder function from K to X .) Moreover a simple heuristic argument (due to L.L. Landau-E.M. Lifshitz [27], §32) yields the following estimate for the number N : Define the maximal mean dissipation of energy ϵ by

$$\begin{aligned}\epsilon &= \limsup_{t\to\infty} \frac{\nu}{t}\int_0^t \frac{1}{\mathrm{vol}\Omega}\|u(\tau)\|^2 d\tau \\ &= \sup \limsup_{t\to\infty} \frac{\nu}{t}\int_0^t \frac{1}{\mathrm{vol}\Omega}\int_\Omega |\nabla u(x,\tau)|^2 dx\, d\tau\end{aligned} \tag{4.1}$$

where the supremum is taken over all weak solutions $u(\cdot)$ on $(-\infty,\infty)$ such that $u(t) \in X$, $\forall t$. The Kolmogorov dissipation length ℓ_d is the only length of the form $\epsilon^\alpha \nu^\beta$, i.e. $\ell_d = (\nu^3/\epsilon)^{1/4}$. It is assumed that the turbulent eddies of linear size $< \ell_d$ are quickly killed by the viscosity of the fluid. Thus a grid with mesh $\sim \ell_d$ will be able to monitor the longer-lasting eddies. Therefore

$$N \sim c_3\left[\frac{\ell_o}{\ell_d}\right]^n \quad \text{where} \quad \ell_o \sim (\mathrm{vol}\Omega)^{1/n} , \tag{4.2}$$

and $n = 2,3$ is the dimension of Ω . The idea to connect the estimate (4.2) with that for $d_M(X)$ was first presented in [28] (see also [24] and [31]). It is a remarkable fact that for $n = 2$ one has

$$d_M(X) \le c_4\left[\frac{\ell_o}{\ell_d}\right]^2 .$$

However the following question is still open.

Question 2. Is the formula

$$d_M(X) \leq c_5\left[\frac{\ell_o}{\ell_d}\right]^3$$

true in the 3D case?

What is known now is the following: Assume that $X \subset V$ and define $\ell_d' = (\nu/\epsilon')^{1/4}$, with

$$\epsilon' = \sup \limsup_{t\to\infty} \nu\left[\frac{1}{t}\int_o^t \frac{1}{\mathrm{vol}\Omega}\int_\Omega |\nabla u(x,\tau)|^{5/2} dx\, d\tau\right]^{4/5} \qquad (4.2)'$$

where the supremum is taken over the same family as in (4.1). Then

$$d_M(X) \leq c_6\left[\frac{\ell_o}{\ell_d'}\right]^3$$

(see [24], §4.5). The odd power in (4.2)$'$ is the best that can be obtained now using the methods of [26], [25], [31] and the Lieb-Thirring Sobolev-type inequality [32].

The physical meaning of the estimate of $d_M(X)$ leads to another open question. Indeed introducing the Reynolds number

$$Re = \frac{1}{\nu\ell_o} \sup \limsup_{t\to\infty} \left[\frac{1}{t}\int_o^t \frac{|u(\tau)|^2}{\mathrm{vol}\Omega}\, d\tau\right]^{1/2}$$

$$= \frac{1}{\nu\ell_o} \sup \limsup_{t\to\infty} \left[\frac{1}{t}\int_o^t \frac{1}{\mathrm{vol}\Omega}\int_\Omega |u(x,\tau)|^2 dx\, d\tau\right]^{1/2}$$

where the supremum is again taken over the same family as in (4.1), a classical argument in the conventional theory of turbulence yields that $\ell_o/\ell_d \sim Re^{3/4}$ and therefore $N \sim Re^{9/4}$ (see again [27], §32). Thus Question 2 has the following supplement

Question 3. Is the formula

$$d_M(X) \leq c_6 Re^{9/4} \tag{4.3}$$

true in the 3D case?

The best known result is the following ([24], §4.5): Assume $X \subset V$ and define

$$Re' = \frac{1}{\nu \ell_o} \sup \limsup_{t\to\infty} \left[\frac{1}{t}\int_0^t \frac{1}{\mathrm{vol}\,\Omega}\int_\Omega |u(x,\tau)|^5 \; dx\, d\tau\right]^{1/5}$$

Then

$$d_M(X) \leq c_7 Re'^3 \tag{4.4}$$

Every rigorous result which would bring (4.4) nearer to (4.3) will constitute an important contribution to the connection between the mathematical theory of fluid dynamics and the conventional theory of turbulence.

It is interesting that the estimates for the Hausdorff dimension of the attractors of dissipative partial differential systems, including the Navier-Stokes equations, do not improve upon ours, when expressed in the natural physical quantities introduced above (see for instance [29], [30] or

[15]). Therefore it would be interesting to prove that for the universal attractor X of the 2D Navier-Stokes equations one has

$$d_M(X) \leq c_8 d_H(X) \quad . \tag{4.5}$$

5. INERTIAL MANIFOLDS AND INERTIAL FORMS

An *inertial form* of a dissipative partial differential equation, in particular of an equation of the form (2.4), is an ordinary differential system

$$\frac{dv}{dt} + N_I(v) = 0 \tag{5.1}$$

in an open bounded domain $\mathcal{O}_I$ in $\mathbb{R}^I$ and a map $\phi_I : \mathcal{O}_I \to V$ satisfying the following conditions:

(i) N_I and ϕ_I are Lipschitz functions,

(ii) $\mathcal{O}_I$ is invariant to (5.1) and there exists a compact set $X_I \subset \mathcal{O}_I$ attracting all solutions of (5.1),

(iii) For every solution $v(\cdot)$ of (5.1), $u(\cdot) = \phi_I(v(\cdot))$ is a solution of (2.4),

(iv) Every solution of (2.4) is exponentially attracted in H (or equivalently in V) to $\phi_I(\mathcal{O}_I)$.

An *inertial manifold* of (2.4) is a Lipschitz manifold which is of the form $\phi_I(\mathcal{O}_I)$ for an appropriate inertial form. Also the existence (resp. determination) of an inertial form provides a good theoretical (resp. practical) base for the numerical simulation of the long time behaviour of a dissipative partial differential equation, provided the dimension I is not too large. It is also clear that the universal attractor X lies on any inertial manifold or more generally that $\phi_I(X_I) \supset X$ for any inertial form. Thus I must be $\geq d_M(X)$. It is also clear that the existence of an inertial form will constitute a sound mathematical framework for the dynamical systems approach to turbulence. Therefore a first question to be asked is:

Question 4. Which equations of the form (2.4) have inertial forms?

Presently the best answers to this question seem to be of the following kind:

Theorem 2. There exist $\delta, \eta, \mu \in (0,1]$ depending on the constants α, β, γ appearing in (2.6)-(2.6)$'$ such that if

$$\mu\nu^{\eta}(\Lambda_{m+1}^{\delta} - \Lambda_m^{\delta}) \geq 1 \quad , \tag{5.2}$$

where $0 < \Lambda_1 \leq \Lambda_2 \leq \ldots$ are the distinct eigenvalues of A , then (2.4) has an inertial manifold.

For a hint on the proof of this type of results we refer to [33], [34].

It can be shown that the hypotheses of the theorem apply to many interesting partial differential equations. In particular this is true for the Kuramoto-Sivashinsky equation

$$u_t + u_{xxxx} + u_{xx} + uu_x = 0 \quad , \tag{5.3}$$

where u is odd and L -periodic, $L > 0$. It was shown that (5.2) has inertial manifolds of dimension $I \leq c_9 L^{3.5}$ [34]. However for the universal attractor X of (5.3) one has $d_M(X) \leq c_{10} L^{1.5}$ [35]. These results leave the following question still open:

Question 5. Does the Kuramoto-Sivashinsky equation (5.3) satisfy the condition

$$c_{11}L \le d_M(X) \le I \le c_{12}L \quad , \tag{5.4}$$

where I stands for the dimension of an appropriate inertial manifold of (5.3)?

A conjecture due to Y. Pomeau based on some strong numerical evidence suggests that (5.4) is true.

Theorem 2 applies also to reaction-diffusion equations with no restriction on the diffusion coefficient ν. Remarkably, an explicit inertial form for systems of such equations with diffusion coefficients restricted to an adequate range was given earlier in [36]. However Theorem 2 does not apply to the Navier-Stokes equations, even if $n = 2$. Indeed in this case $\delta = \frac{1}{2}$ and the condition (5.2) is never satisfied if ν is small enough. Therefore we finish with a particular version of Question 4, which we hope will turn out to be true.

QUESTION 6. Do the 2D Navier-Stokes equations always have inertial manifolds?

REFERENCES

[1] O. Reynolds, Phil. Trans. Roy. Soc. London, 186(1894), 123.

[2] G.I. Taylor, Proc. Roy. Soc. A, 151(1935), 421 and 156(1936), 307.

[3] T. von Kármán, J. Aero. Sci., 4(1937), 131.

[4] A.N. Kolmogorov, C.R. Acad. Sci. U.R.S.S., 30(1941), 301.

[5] J. Leray, Acta Math., 63(1934), 193.

[6] B. Mandelbrot, The fractal geometry of Nature, Freeman & Co., New York (1983).

[7] L.L. Landau, Doklady Akad. Nauk, 44(1944), 311.

[8] E. Hopf, Comm. Pure Appl. Math., 1(1948), 303.

[9] D. Ruelle-R. Takens, Comm. Math. Phys., 20(1971), 167 and 23(1971), 343.

[10] V. Scheffer, Comm. Math. Phys., 55(1977), 97.

[11] L. Caffarelli-R.V. Kohn-L. Nirenberg, Comm. Math. Pure Appl., 35(1982), 771.

[12] J.P. Kahane, C.R. Acad. Sci. Paris, 278A(1974), 621.

[13] J. Mallet-Paret, Journ. Diff. Eq., 22(1976), 331.

[14] R. Temam, Navier-Stokes equations and nonlinear functional analysis, CBMS-NSF Reg. Conf. Sem. in Appl. Math., SIAM, Philadelphia (1983).

[15] A.V. Babin-M.I. Vishik, Uspehi Mat. Nauk, 38(4)(1983), 133.

[16] D. Ruelle, Bull. Amer. Math. Soc., 5(1981), 29.

[17] E.N. Lorenz, Journ. Atmosph. Sci., 20(1963), 130.

[18] P. Bernard-T. Ratiu (edts), Turbulence Seminar, Lecture Notes in Math., #615, Springer, New York (1977).

[19] G.I. Barenblatt-G. Ioos-C.C. Joseph (edts), Nonlinear Dynamics and Turbulence, Putnam, Boston (1983).

[20] H.L. Swinney-J.P. Gollub (edts), Hydrodynamic instabilities and the Transition to Turbulence (2nd Ed.), Topics in Applied Phys., 45, Springer, New York (1985).

[21] P. Bergé-Y. Pomeau-Chr. Vidal, L'ordre dans le chaos, Hermann, Paris (1984).

[22] J.L. Lions, Quelques méthodes de résolutions des problèmes aux limites non linéares, Dunod-Gauthier--Villars, Paris (1969).

[23] R. Temam, Navier-Stokes equations, Theory and Numerical Analysis (2nd Ed.), North Holland, Amsterdam (1979).

[24] P. Constantin-C. Foias-R. Temam, Attractors representing turbulent flows, Memoirs of the AMS, 53, #314(1985).

[25] C. Foias-R. Temam, Journ. Math. Pure Appl., 58(1979), 339.

[26] P. Constantin-C. Foias, Comm. Pure Appl. Math., 38(1985), 1.

[27] L.L. Landau-E.M. Lifshitz, Fluid Mechanics, Pergamon, New York (1959).

[28] P. Constantin-C. Foias-O.P. Manley-R. Temam, C.R. Acad. Sci. Paris, I, 297(1983), 599.

[29] D. Ruelle, Comm. Math. Phys., 87(1982), 287.

[30] E. Lieb, Comm. Math. Phys., 92(1984), 473.

[31] P. Constantin-C. Foias-O.P. Manley-R. Temam, J. Fluid Mech., 150(1985), 427.

[32] E. Lieb-W. Thirring, Studies in Math. Physics: Essays in Honor of V. Bargman (E. Lieb-E. Simon-A.S. Wightman, edts), Princeton Univ. Press, Princeton (1976), 269.

[33] C. Foias-G.R. Sell-R. Temam, C.R. Acad. Sci. Paris, I, 301(1985), 139.

[34] C. Foias-B. Nicolaenko-G.R. Sell-R. Temam, C.R. Acad. Sci. Paris, I, 301(1985), 285.

[35] B. Nicolaenko-B. Scheurer-R. Temam, Physica D, 16(1985), 155.

[36] E. Conway-D. Hoff-J. Smoller, SIAM J. Appl. Math., 35(1978),1.

Mathematics Department
Indiana University
Bloomington, IN 47405

Université de Paris-Sud
Mathématiques
Bât 425
91405 Orsay Cédex FRANCE

Blow-Up of Solutions of Nonlinear Evolution Equations

Avner Friedman

INTRODUCTION.

Consider a nonlinear evolution equation

$$\frac{\partial u}{\partial t} - A\left(u, D_x u, \ldots, D_x^m u\right) = 0 \qquad (x \in \Omega, t > 0) \tag{0.1}$$

with initial condition

$$u(x,0) = \phi(x) \qquad (x \in \Omega) \tag{0.2}$$

where $u = u(x,t)$ is either a scalar or vector function and Ω is a domain in $\mathbb{R}^N$. If $\Omega \neq \mathbb{R}^N$ then we also prescribe boundary conditions on $\partial\Omega$ for $t > 0$. We are interested in the situation where a global classical solution does not exist for all $t > 0$. The basic problem is then to study the nature of the singularities of $u(x,t)$ at the time when the solution ceases to be a classical solution.

We shall review some recent work. The results and methods depend upon the particular type of the linearized equation and upon the nonlinear terms.

DIRECTIONS IN PARTIAL
DIFFERENTIAL EQUATIONS

ISBN 0-12-195255-X

§1. Nonlinear parabolic equations.

Consider the parabolic equation

$$
\begin{aligned}
&u_t - \Delta u = f(u) && (x \in \Omega, t > 0) \\
&u(x,o) = \phi(x) && (x \in \Omega) \qquad (1.1)\\
&u(x,t) = 0 && (x \in \partial\Omega, t > 0)
\end{aligned}
$$

where Ω is a bounded domain in $\mathbb{R}^N$ with $C^{1,\alpha}$ boundary and $\phi \geqslant 0$, $\phi = 0$ on $\partial\Omega$. We assume that $f \in C^2[0,\infty)$ and

$$
\begin{aligned}
&f(s) > 0 \text{ if } s > 0 \\
&\int^{\infty} \frac{ds}{f(s)} < \infty .
\end{aligned} \qquad (1.2)
$$

Then the solution u is positive and therefore either it exists for all $t > 0$ or else there is a $T < \infty$ such that u exists for all $t < T$ and

$$\max_{x \in \overline{\Omega}} u(x,t_n) \longrightarrow \infty \quad \text{for a sequence} \quad t_n \to T . \qquad (1.3)$$

A variety of sufficient conditions are known which ensure the blow-up (1.3); see [1,17,18,25] and the references given there. One standard argument goes as follows:

If (ψ,λ_1) is the principal eigensolution of Δ in Ω then the function $g(t) = \int_\Omega u(x,t)\psi(x)dx$ satisfies

$$g'(t) \geqq -\lambda_1 g(t) + f(g(t)) \qquad (1.4)$$

by (1.1) and Jensen's inequality (assuming f to be convex and ψ normalized : $\psi > 0$, $\int_\Omega \psi = 1$). Consequently if $g(0)$ is large enough then $g(t)$ must blow-up in finite time.

Suppose $f(0) > 0$ and denote by δ_* the g.l.b. of the spectral points of $\Delta u + \delta f(u)$ in Ω (under zero boundary conditions). Then if $\delta > \delta_*$, $f' > 0$ and f/f' is concave, the solution of (1.1) blows-up in finite time, for <u>any</u> $\phi \not\equiv 0$.

This result of Bellout [2] includes the cases

$$f(u) = e^u , \quad f(u) = (C+u)^p \qquad (C>0, p>1); \tag{1.5}$$

a similar result was established earlier by Lacey [17], but it includes (1.5) only if $N \leqslant 10$ and $N \leqslant 6$ respectively. The proof in [17] develops an extension of the argument involving (1.4), whereas the proof in [2] is based on a comparison argument.

A point x is said to belong to the <u>blow-up set</u> S of u if there exists a sequence (x_n, t_n) such that

$$x_n \to x, \quad t_n \to T, \quad u(x_n, t_n) \to \infty \text{ if } n \to \infty .$$

Clearly S is nonempty and closed in $\bar{\Omega}$.

Consider the case $N = 1$, $\Omega = \{-a < x < a\}$ and assume that

$$\begin{aligned} \phi'(x) &\geqslant 0 \quad \text{if} \quad -a < x < \gamma \\ \phi'(x) &\leqslant 0 \quad \text{if} \quad \gamma \leqslant x < a \end{aligned} \tag{1.6}$$

for some $-a < \gamma < a$; furthermore, if $\phi(x)$ is not symmetric about $x = 0$ then we also assume that

$$f(\phi) + \phi'' \geqslant 0 , \tag{1.7}$$

which means that $u_t(x,0) \geqslant 0$ (and then $u_t \geqslant 0$ by the maximum principle).

<u>Theorem 1.1</u>. <u>Under the assumptions</u> (1.6), (1.7) <u>the blow-up set consists of a single point</u>.

Here f is assumed to be any convex function and, in case $f(0) = 0$, it satisfies an additional (rather weak) technical condition; the examples in (1.5) are included.

Theorem 1.1 is due to Friedman and Mcleod [7]; in the symmetric case, and under additional restrictions on ϕ and f, the theorem was earlier proved by Weissler [25].

The proof in [7] is based on estimates from below on $|u_x|$, in the regions $\{x < s(t)\}$ and $\{x > s(t)\}$ where $x = s(t)$ is a point of maximum of the function $x \to u(x,t)$. From these estimates it follows that u cannot blow-up to the right of $A \equiv \overline{\lim}_{t\to T} s(t)$ and to the left of $B \equiv \underline{\lim}_{t\to T} s(t)$. Next one establishes:

Lemma 1.2. The blow-up set is a compact subset of Ω.

That means that $A < a$ and $B > -a$. Finally one proves that $A = B$ so that $x_0 = \lim_{t\to T} s(t)$ is the single blow-up point.

Lemma 1.2 is actually valid for any N-dimensional convex domain Ω and this fact is used, in [7], in order to derive an upper bound on the rate of blow-up of u:

$$u(x,t) \leqslant G^{-1}(\delta(T-t))$$

for some $\delta > 0$, where

$$G(s) = \int_s^\infty \frac{d\tau}{f(\tau)} .$$

In particular, if

$$f(s) = (\mu + s)^p \qquad (\mu \geqslant 0, p > 1) \tag{1.8}$$

then

$$u(x,t) \leqslant C_0(T-t)^{-q} \qquad \left(q = \frac{1}{p-1}\right). \tag{1.9}$$

In the radially symmetric case (1.9) was also established by Weissler [26].

Theorem 1.3. Consider the case (1.8) with $1 < p \leqslant (N+2)/(N-2)$. If x_0 is a blow-up point then

$$u(x,t) \sim A(T-t)^{-q}$$

if $|x-x_0| \leqslant C(T-t)^{1/2}$, $t \to T$ where either $\gamma = 0$ or $\gamma = q^q$.

This result is due to Giga and Kohn [10]. For the proof one takes any blow-up sequence about (x_0,T) (convergence of a subsequence is assured by the estimate (1.9)) and establishes that the limit of any convergent subsequence is unique (if it is nonzero).

In case (1.8) one can show [7] that

$$\underline{\lim}_{t\to T} \|u(\cdot,t)\|_{L^j(\Omega)} = \infty \text{ if } j > \frac{N(p-1)}{2}, \tag{1.10}$$

whereas in the radially symmetric case (with $\phi_r \leq 0$)

$$\overline{\lim}_{t\to T} \|u(\cdot,t)\|_{L^j(\Omega)} < \infty \text{ if } j < \frac{N(p-1)}{2}. \tag{1.11}$$

Notice that (1.11) implies that the blow-up set S has Lebesgue measure zero (in the radially symmetric case with $\phi_r \leq 0$, S consists of a single point, namely $x=0$). It is not known whether (1.11) holds in general, or even whether S always has measure zero. In fact, it is not even known whether the assertion of Theorem 1.1 remains valid for ϕ satisfying (1.7) but not (1.6).

The results from [7] described above are valid also for the boundary condition $\partial_\nu u + \beta u = 0$ where $\beta \geq 0$, $\beta_t \leq 0$; see [7]. Some of the results extend to a system of parabolic equations [6].

§2. Degenerate nonlinear parabolic equations.

The results described in this section are taken from Friedman and Mcleod [8].

Consider

$$u_t = u^2(u_{xx} + u) \qquad (-a < x < a, t > 0) \tag{2.1}$$

$$u(x,0) = \phi(x) \qquad (-a < x < a) \tag{2.2}$$

$$u(\pm a,t) = 0 \qquad (t > 0) \tag{2.3}$$

where $\phi(x) > 0$ if $|x| < a$, $\phi(\pm a) = 0$. This initial-boundary value problem arises in a model for resistive diffusion of force free magnetic field in plasma confined between two walls. We consider solutions which are positive if $-a < x < a$, $t > 0$ and continuous up to the boundary.

Theorem 2.1. (i) *If* $a < \pi/2$ *then there exists a solution for all* $t > 0$ *and* $u(x,t) \leqslant C/\sqrt{t}$ *as* $t \to \infty$. (ii) *If* $a > \pi/2$ *then there exists a unique solution* u *for all* $0 < t < T$, *for some* $T < \infty$, *and the solution blows up as* $t \to T$.

More generally, if Ω is any bounded domain, then the problem

$$\begin{aligned} u_t &= u^2(\Delta u + u) && (x \in \Omega, t > 0) \\ u(x,0) &= \phi(x) && (x \in \Omega) \\ u(x,t) &= 0 && (x \in \partial\Omega, t > 0) \end{aligned} \tag{2.4}$$

has a global solution (bounded by $C/\sqrt{t}$) if $\lambda_1 > 1$ and the solution blows up in finite time if $\lambda_1 < 1$; here λ_1 is the principal eigenvalue of Δ in Ω.

Consider now the problem (2.1) - (2.3) under the conditions:

$$\begin{aligned} &\frac{\pi}{2} < a < \infty \\ &\phi(x) = \phi(-x), \ \phi'(x) < 0 \text{ if } 0 < x < a \\ &\phi'' + \phi \geqslant 0, \end{aligned} \tag{2.5}$$

and denote the blow-up time by T. Notice that $u_x < 0$ if $0 < x < a$ and, consequently,

$$u(0,t) = \max_{|x| \leqslant a} u(x,t) \longrightarrow \infty \text{ if } t \to T.$$

Theorem 2.2. *The blow-up set is precisely the set* $\left\{ -\frac{\pi}{2} \leqslant x \leqslant \frac{\pi}{2} \right\}$; *furthermore,*

$$\varliminf_{t \to T} \frac{u(x,t)}{u(0,t)} \geqslant c(x) > 0 \ \underline{\text{if}} \ 0 < x < \frac{\pi}{2},$$

and

$$\frac{u(0,t)}{\sqrt{T-t}} \longrightarrow \infty \ \underline{\text{if}} \ t \to T. \tag{2.6}$$

It would be interesting to find the precise asymptotic behavior of the function in (2.6); some numerical results are given in [22].

We finally mention that for (2.4), with $\Omega \subset \mathbb{R}^N$ and $\lambda_1 < 1$, the blow-up set S has positive measure and $\lambda_1(G) \leqslant 1$ for any open set $G \supset S$.

§3. Nonlinear wave equations.

Consider the nonlinear wave equation

$$u_t - \Delta u = |u|^p \qquad (x \in \mathbb{R}^N, t > 0) \tag{3.1}$$

where $p > 1$, with the Cauchy data

$$u(x,0) = f(x),\ u_t(x,0) = g(x) \qquad (x \in \mathbb{R}^N)\ . \tag{3.2}$$

It is well known that classical solutions blow-up if the Cauchy data are "large" in some sense; see [11,13,16,19] and the references given there. If $N = 3$ and $1 < p < 1+\sqrt{2}$ then for _any_ Cauchy data with compact support the solution blows up in finite time, unless $u \equiv 0$; this result is due to John [14].

We consider from now on the case where smooth solutions do not exist for all $t > 0$.

Theorem 3.1. _Suppose_ $N = 1$ _and_ $f,g \in C^{3,1}$. _Then there exists a finite valued function_ $\phi(x)$ _and a smooth solution_ $u(x,t)$ _of_ (3.1), (3.2) _in_ $\{t < \phi(x), x \in \mathbb{R}^1\}$; _further_, $\phi \in C^1$, $|\nabla\phi(x)| < 1$, _and_

$$d^q(x,t)u(x,t) \longrightarrow \left[\frac{A(1-\alpha^2)}{1+\alpha^2}\right]^{1/(p-1)} \qquad \left(A = \frac{2(p+1)}{(p-1)^2}\right)$$

_if_ $d(x,t) \equiv \mathrm{dist}((x,t),\{t=\phi\}) \to 0$, $x \to x^0$ _where_ $\alpha = |\nabla\phi(x^0)|$, $q = 2/(p-1)$, _and_

$$\frac{u_t^2 - |\nabla u|^2}{u^{p+1}} \to \frac{2}{p+1}\ , \qquad \frac{u_t^2}{u^{p+1}} \to \frac{2}{(p+1)(1-\alpha^2)}\ , \qquad \frac{u_{tt}}{u^p} \to \frac{1}{1-\alpha^2}\ .$$

The pair (u,ϕ) _is uniquely determined_.

This theorem is due to Caffarelli and Friedman [5]. The proof's outline is as follows:

Using the integral representation of u one shows that $u_t \pm u_x \geqslant -C$. This enables us to define a curve $t = \phi(x)$ such that the solution exists if $t < \phi(x)$ and blows up to $+\infty$ if $t \uparrow \phi(x)$; ϕ is Lipschitz continuous with coefficient 1. Next one estimates u_t and u_{tt} (using again integral representation) and thereby derives estimates from above and below (in terms of u or d) near the blow-up curve $\{t=\phi\}$. The next step involves working with blow-up sequences

$$u_\lambda(x) = \lambda^q u(x_0 + \lambda x, t_0 + \lambda t) \qquad (\lambda \downarrow 0)$$

where $t_0 = \phi(x_0)$, and establishing that there is a unique limit, namely, $C_\alpha\left((\alpha x-t)^+\right)^q$ (C_α constant). Here we use the preceding estimates in order to establish the existence of limits of sequences u_{λ_n} and also to show that they are convex functions. The analysis involved is somewhat in the spirit of elliptic methods in free boundary problems.

Having established the linearity of the blow-up limit curve it follows that $\phi'(x_0)$ exists; the continuity of $\phi'(x)$ now easily follows.

We next formulate, somewhat imprecisely, an extension of Theorem 3.1 to $N \leqslant 3$:

<u>Theorem 3.2</u>. <u>Theorem</u> 3.1 <u>extends to the case</u> $N \leqslant 3$ <u>provided</u> f <u>and</u> g <u>are such that the solution</u> $\tilde{u}$ <u>of</u> $\tilde{u}_t = \Delta\tilde{u}$ <u>with data</u> (3.2) <u>satisfies</u>

$$\tilde{u}_t > (1 + \varepsilon_0)|\nabla_x \tilde{u}|$$

<u>for some</u> $\varepsilon_0 > 0$.

This result is due to Caffarelli and Friedman [4]; the precise formulation requires some additional minor technical assumptions on f,g.

It would be interesting to establish further regularity of $\phi(x)$.

§4. The Hamilton-Jacobi equation

The results of this section are based on recent work by Friedman and Souganidis [9].

Consider viscosity solutions of a Hamilton-Jacobi equation

$$\begin{aligned} &u_t + H(D_x u) = |u|^p && (x \in \mathbb{R}^N, t > 0) \\ &u(x,0) = f(x) && (x \in \mathbb{R}^N) \end{aligned} \tag{4.1}$$

where $p > 1$ and $H(w) \to \infty$ if $|w| \to \infty$. Assume that $f \geqslant 0$ and

$$f^p - H(D_x f) \geqslant \varepsilon_0(1 + |D_x f|), \qquad \varepsilon_0 > 0. \tag{4.2}$$

Then there exists a surface $t = \phi(x)$ with ϕ Lipschitz continuous such that there exists a viscosity solution u of (4.1) in $\{t < \phi(x)\}$ and it blows up to $+\infty$ as $d(x,t) \to 0$, where $d(x,t) = \text{dist } (x,t),\{t=\phi\}$.

If $0 \leqslant H(w) \leqslant C(|w| + 1)$ then the blow-up rate is given by

$$0 < c \leqslant d^q(x,t)u(x,t) \leqslant C < \infty \ , \quad q = \frac{1}{p-1} \ . \tag{4.3}$$

If however $H(w) = |w|^2$, say, and $p > 2$, then the blow-up rate depends on which of the two terms u_t and $|D_x u|^2$ is dominant. In case $N = 1$, $\phi(x) = \phi(-x)$, $\phi'(x) < 0$, it can be shown that (4.3) holds for $x \neq 0$, $q = 2/(p-2)$ provided ϕ satisfies some additional technical conditions. On the other hand, without assuming these conditions the assertion is not generally true: counterexamples show that in the blow-up estimates (4.3) q may vary from $1/(p-1)$ in one section of $\{t=\phi\}$ to $q = 2/(p-2)$ in another section of $\{t=\phi\}$.

In general ϕ is not continuously differentiable. If the condition (4.2) is relaxed by taking $\varepsilon_0 = 0$ then the blow-up surface does not generally have the form $t = \phi(x)$; it may consist, for instance, of two rays (when $N = 1$): the positive t-axis and the positive x-axis.

§5. Other equations

Consider the nonlinear Schrödinger equation

$$i\phi_t + \Delta\phi + |\phi|^{2\sigma}\phi = 0 \qquad (x \in \mathbb{R}^N, t > 0) \qquad (5.1)$$

$$\phi(x,0) = \phi_0(x) \qquad (x \in \mathbb{R}^N). \qquad (5.2)$$

If $\sigma \geq 2/N$ then there exist initial data $\phi_0 \in H^1$ for which the solution $\phi(x,t)$ blows up in finite time [12], i.e.,

$$\lim_{t\to T} \int_{\mathbb{R}^N} |\nabla\phi(x,t)|^2 dx = \infty$$

for some $T < \infty$.

A function $\phi = R(x)e^{it}$ is a solution of (5.1) if and only if

$$\Delta R - R - |R|^{2\sigma}R = 0 . \qquad (5.3)$$

If $\sigma = 2/N$ then there exists a solution R of (5.3) which is positive, radial and monotonically decreasing: if $1 \leq N \leq 3$ then the solution is known to be unique (see [3][21]). R is called the <u>ground state</u> solution. If $\sigma = 2/N$ and $\|\phi_0\| < \|R\|$ $\left(\|\cdot\| = \|\cdot\|_{L^2(\mathbb{R}^N)}\right)$ then there exists a global solution of (5.1), (5.2) [23]; if however $\|\phi_0\| = \|R\|$ then the solution may blow-up in finite time [24].

Consider the case $\sigma = 2/N$, $1 \leq N \leq 3$, $\|\phi_0\| = \|R\|$ and assume that the solution blows up at finite time T. Then, up to translations in space and phase,

$$\phi(x,t) \sim \frac{1}{(\lambda(t))^{N/2}} R\left(\frac{x}{\lambda(t)}\right) \text{ as } t \to T \qquad (5.4)$$

where $\lambda(t) = \|\nabla R\|/\|\nabla\phi(t)\|$; this is a recent result by M. I. Weinstein [24]. It is not known what is the behavior of $\phi(x,t)$ near the blow-up time in case $\|\phi_0\| > \|R\|$.

The proof of (5.4), in [24], depends upon showing that if

$$S_\lambda\phi(x,t) = \lambda^{1/\sigma}\phi(\lambda x,t)$$

then the scaled functions

$$S_{\lambda(t_k)}\phi(\cdot + y(t_k), t_k)e^{i\gamma(t_k)}$$

converge strongly in H^1 to the ground state R, for any sequences $t_k \to T$, $y(t_k) \in \mathbb{R}^N$, $\gamma(t_k) \in \mathbb{R}^1$.

We finally consider the question of formation of singularities in the compressible Euler equations

$$\begin{aligned}
&\rho_t + \nabla\cdot\rho u = 0 \\
&\rho(u_t + u\cdot\nabla u) + \nabla p = 0 \\
&S_t + u\cdot\nabla S = 0 \\
&p = A\rho^\gamma e^S \qquad (A > 0, \gamma > 1)
\end{aligned} \tag{5.5}$$

with initial data

$$\begin{aligned}
&\rho(x,0) = \rho^0(x) > 0, && \rho^0(x) = \bar{\rho} \text{ if } |x| > R \\
&u(x,0) = u^0(x), && u^0(x) = 0 \text{ if } |x| > R \\
&S(x,0) = S^0(x), && S^0(x) = \bar{S} \text{ if } |x| > R.
\end{aligned} \tag{5.6}$$

Set

$$m(t) = \int_{\mathbb{R}^3} (\rho(x,t) - \bar{\rho})dx ,$$

$$\eta(t) = \int_{\mathbb{R}^3} \left(\rho(x,t)\ e^{S(x,t)/\gamma} - \bar{\rho}e^{\bar{S}/\gamma}\right) dx,$$

$$F(t) = \int_{\mathbb{R}^3} x\cdot\rho u(x,t)dx.$$

It is well known [15] that there exists a local C^1 solution of (5.5), (5.6). However, a global smooth solution does not exist in general. Recently Sideris [20] has proved that if

$$m(0) \geqslant 0,\ \eta(0) \geqslant 0,\ F(0) \geqslant \alpha\sigma R^4 \max \rho^0(x) \qquad \left(\alpha = \frac{16\pi}{3}\right),$$

then the life span of C^1 solutions is finite. The proof is based on deriving a differential inequality

$$F' \geqslant \beta(t)F^2$$

which does not have a global solution.

REFERENCES

[1] J. W. Bebernes and D. R. Kassoy, A mathematical analysis of blow-up for thermal reactions - the spatially non-homogeneous case, SIAM J. Appl. Math., 40 (1981), 476-484.

[2] H. Bellout, A criterion for blow-up solutions to semi-linear heat equations, to appear.

[3] H. Berestycki and P. L. Lions, Nonlinear scalar field equations I, II, Archive Rat. Mech. Anal., 82 (1983), 313-376.

[4] L. A. Caffarelli and A. Friedman, The blow-up boundary for nonlinear wave equations, to appear.

[5] L. A. Caffarelli and A. Friedman, Differentiability of the blow-up curve for one dimensional nonlinear wave equations, to appear.

[6] A. Friedman and Y. Giga, A single point blow-up for solutions of a semilinear parabolic system (in preparation).

[7] A. Friedman and B. Mcleod, Blow-up of positive solutions of semilinear heat equations, Indiana Univ. Math. J., 34 (1985), 425-447.

[8] A. Friedman and B. Mcleod, Blow-up of solutions of non-linear degenerate parabolic equations, to appear.

[9] A. Friedman and P. E. Souganidis, Blow-up of solutions of Hamilton-Jacobi equations, Comm. in P.D.E., to appear.

[10] Y. Giga and R. V. Kohn, Asymptotically self-similar blow-up of semilinear heat equations, Comm. Pure Appl. Math., 38 (1985), 297-319.

[11] R. T. Glassey, Blow-up theorems for nonlinear wave equations, Math. Z., 132 (1973), 183-203.

[12] R. T. Glassey, On the blowing up of solutions to the Cauchy problem for the nonlinear Schrödinger equation, J. Math. Phys., 18 (1977), 1794-1797.

[13] R. T. Glassey, Finite-time blow-up for solutions of non-linear wave equations, Math. Z., 177 (1981), 323-340.

[14] F. John, Blow-up of solutions of nonlinear wave equations in three space dimensions, Manusc. Math., 28 (1979), 235-268.

[15] T. Kato, The Cauchy problem for quasilinear symmetric systems, Arch. Rat. Mech. Anal., 58 (1975), 181-205.

[16] T. Kato, Blow-up solutions of some nonlinear hyperbolic equations, Comm. Pure Appl. Math., 32 (1980), 501-505.

[17] A. A. Lacey, Mathematical analysis of thermal runaway for spatially inhomogeneous reactions, SIAM J. Appl. Math., 43 (1983), 1350-1366.

[18] H. A. Levine, Some nonexistence and stability theorems of formally parabolic equations of the form $Pu_t = -Au + F(u)$, Arch. Rat. Mech. Anal., 51 (1973), 371-386.

[19] H. A. Levine, Instability and nonexistence of global solutions to nonlinear wave equations of the form $Pu_{tt} = -Au + F(u)$, Trans. Amer. Math. Soc., 192 (1974), 1-21.

[20] T. C. Sideris, Formation of singularities in three-dimensional compressible fluids, to appear.

[21] W. A. Strauss, Existence of solitary waves in higher dimensions, Commun. Math. Phys., 55 (1977), 149-162.

[22] P. A. Walterson, Force-free magnetic evolution in the reversal-field pinch, Thesis, Cambridge University, 1984.

[23] M. I. Weinstein, Nonlinear Schrödinger equations and sharp interpolation estimates, Commun. Math. Phys., 87 (1983), 567-576.

[24] M. I. Weinstein, On the structure and formation of singularities in solutions to nonlinear dispersive evolution equations, to appear.

[25] F. B. Weissler, Single point blow-up of semilinear initial value problems, J. Diff. Eqs., 55 (1984), 204-224.

[26] F. B. Weissler, An L^{∞} blow-up estimate for a nonlinear heat equation, Comm. Pure Appl. Math., 38 (1985), 291-295.

The author is partially supported by National Science Foundation Grant DMS-8420896.

Purdue University
Department of Mathematics
West Lafayette, Indiana 47907

Coherence and Chaos in the Kuramoto-Velarde Equation

James M. Hyman and Basil Nicolaenko

I. INTRODUCTION

In the past decade, research into the behavior of finite dimensional dynamical systems has deepened our understanding of the transition to chaos in dissipative physical models. Underlying this explosive research field is the belief that transitions to chaos are generic, low-dimensional phenomena, even for continuum fluid flows governed by an infinite number of degrees of freedom. This is the essence of Feigenbaum's universality theory [19].

Yet, the fundamental problem remains open: are these transitions to dynamical chaos a route to fully developed turbulence in complex fluid flows [2-5, 9-11, 33-34]? A growing body of carefully controlled experiments suggests this is the case [7, 32], although when critical parameters such as the Reynolds number reach even moderate values, the physical detection of a small set of degrees of freedom is quickly blurred by statistical experimental noise [30].

Current computer calculations are unable to accurately simulate the full route to three-dimensional fluid turbulence. The next generation of supercomputers will have enough brute force to allow us to glimpse at the details of the transition to turbulence in the Navier-Stokes equations. Before they are available, we can study turbulent phenomena which can be modeled by one- and two-dimensional scalar partial differential equations (PDEs). In this paper, we

ISBN 0-12-195255-X

focus on PDEs that model localized patterns and structures appearing on interfaces between complex flows. They occur in quasi-planar flame fronts [4], thin viscous fluid films flowing over inclined planes, and the dendritic phase change fronts in binary alloy mixtures [31]. The solution of some of these models is indistinguishable from the solution of a finite-dimensional dynamical system.

In these models, as a critical physical parameter is varied, a simple laminar solution destabilizes. The destabilization is heralded by the cohesive organization of cells and patterns (often hexagonal) on the moving front or interface. The turbulence, localized on the interface, is dominated by fluctuations in the normal direction. As the critical parameter is increased, the spatial cells remain coherent, yet temporal behavior becomes chaotic. This behavior has been observed in flame-sheet experiments during the transition to fully turbulent flames [1]. The coexistence of spatial coherence with temporal chaos in these experiments makes them superb candidates for mathematical and computational testing of the link between deterministic chaos [19] and turbulence.

Many interfaces with localized turbulence, including flames, can be modeled by the simple Kuramoto-Sivashinsky (K-S) PDE [26-29, 31]. This equation accurately accounts for the thermodiffusive and convective mechanism of flow-field coupling across an interface before turbulence breaks away from the interface and reaches deeply into the fluid. In one space dimension, the K-S equation modeling a small perturbation $u(x,t)$ of a metastable planar front or interface is

$$u_t + \nu u_{xxxx} + u_{xx} + \tfrac{1}{2}(u_x)^2 = 0 \quad , \quad (x,t) \in R^1 \times R_+ \quad ,$$
$$u(x,0) = u_0(x) \quad , \quad u(x + L) = u(x,t) \quad . \tag{1.1}$$

Here the subscripts indicate partial differentiation, ν is a positive fourth-order viscosity, and u_0 is L-periodic; L being the size of a typical pattern scale. The natural bifurcation parameter is the renormalized dimensionless parameter $\tilde{L} = L/(2\pi\sqrt{\bar{\nu}})$; $[\tilde{L}]$ is also the number of linearly unstable Fourier modes.

A related model is the Kuramoto-Velarde (K-V) equation [17]:

$$u_t + \nu u_{xxxx} + u_{xx} + \beta u$$

$$+ \gamma(u_x)^2 + \delta(uu_x)_x - \frac{\gamma}{L}\int_0^L (u_x)^2\, dx = 0 \quad , \qquad (1.2)$$

$$\int_0^L u(x,t)\, dx \equiv 0 \quad ,$$

where β,γ,δ are positive, $\beta \ll 1$, and with periodic boundary conditions and initial conditions with zero mean. The K-V equation models Benard-Marangoni cells that occur when there is large surface tension on the interface [35, 36] in a microgravity environment [16, 17]. This situation arises in crystal growth experiments aboard an orbiting space station. Although the free interface is metastable with respect to small perturbations, the nonlinearity $\delta(uu_x)_x$, not present in the K-S equation, models pressure destabilization effects striving to rupture the interface.

Our computer simulations of the K-S equation [24] demonstrated an uncanny intermittent, low-dimensional behavior for the values of the bifurcation parameter beyond the point where the transition to chaos had occurred. Some canonical mechanisms for onset of chaos [19] in classical dynamical systems are seen amidst complex and lengthy turbulent time series. In this article, we report on a similar systematic study of the transition to chaos for the K-V equation.

If the eikonal nonlinearity term u_x^2 in (1.2) is removed ($\gamma = 0$, $\beta \geqq 0$), then the solution of the K-V equation and its gradients blow up in finite time. We have verified this numerically and demonstrated it using classical methods [Ladyshenskaia]. Furthermore, in our numerical studies we found that the convective term u_x^2 not only controls blowup in (2.6), but also generates <u>chaotic dynamics qualitatively similar to those of K-S</u>. In this paper we systematically search for classical dynamical systems bifurcations and for multiple basins of attractions for the K-V model. As in the K-S model [24], we have uncovered a rich solution structure with multiple internal attractors having interlaced basins of attraction.

When we solve chaotic PDEs, the reliability of the long-term, time-dependent behavior is always in question until a comprehensive error analysis is performed. In our computer experiments, we found that the calculated solutions could be extremely sensitive to the numerical accuracy. Nonconverged numerical solutions of the K-S and K-V equations can occur in regimes we are interested in if the time integration errors are greater than 10^{-6} per unit timestep. In fact, small effects of the order of 10^{-6} in the energy for some sensitive Fourier modes can critically impact on the nonlinear dynamics. In the calculations, we used discrete Fourier transform pseudospectral approximations to the spatial derivatives [22] on grids ranging from 64 to 256 mesh points in single precision (14 digits) on a Cray XMP computer. The solution was integrated in time using a variable order, variable timestep backward differentiation method [21] that retained an absolute error tolerance between 10^{-6} and 10^{-10} per unit time. The runs presented here took between 10^4 and 10^5 time-steps. The implicit equation was solved on each timestep with a quasi-Newton iterative algorithm. Because these equations were not solved exactly, a symmetry-breaking perturbation was introduced into the calculation. Many of the current calculations use an approximate solution operator (ASO) based on finite Fourier transforms and an exponential trapezoidal rule [23]. The ASO methodology incorporates analytic information on the behavior of the solution into the numerical approximation.

A typical example of the extreme numerical sensitivity of the numerical solutions to the K-S and K-V equations is the disappearance of homoclinic orbits if the precision is too low. The saddle-type hyperbolic fixed points degenerate into numerically stable fixed points with an artificial basin of attraction the size of the error control. Because of the artificially stable fixed points, our numerical results of the K-S and K-V equations differ from some of the previously published simulations with only modest control over time integration errors.

II. OVERVIEW OF COMPUTATIONAL SIMULATIONS AND THEORETICAL RESULTS

In our calculations, we normalized the K-S equation to an interval of length 2π, set the damping parameter to the original value derived by Sivashinsky ($\nu = 4$), and introduced the bifurcation parameter $\alpha = 4\tilde{L}^2 = L^2/4\pi^2$:

$$u_t + 4u_{xxxx} + \alpha\,[u_{xx} + \tfrac{1}{2}(u_x)^2] = 0 \quad , \quad 0 \le x \le 2\pi \quad , \tag{2.1}$$

$$u(x + 2\pi, t) = u(x,t) \quad , \quad u(x,0) = u_o(x) \quad .$$

This equation is equivalent to Eq. (1.1) with a different time scaling.

The mean value of the solution to Eq. (2.1)

$$m(t) = \frac{1}{2\pi} \int_0^{2\pi} u(x,t)dx \tag{2.2}$$

satisfies the drift equation

$$\dot{m}(t) = \frac{-\alpha}{4\pi} \int_0^{2\pi} (u_x)^2 \, dx \quad . \tag{2.3}$$

To normalize this drift to zero, we numerically solved the drift-free K-S equation for

$$v(x,t) = u(x,t) - m(t) \quad . \tag{2.4}$$

That is,

$$v_t + 4v_{xxxx} + \alpha[v_{xx} + \tfrac{1}{2}(v_x)^2] + \dot{m}(t) = 0 \quad . \tag{2.5}$$

We also normalized the K-V equation on $[0,2\pi]$ and set $\nu = 4$, $\gamma = \frac{1}{2}$, $\delta = 1$ (comparable nonlinearities). In most situations $\beta << 1$ and, therefore, for simplicity and to allow a systematic comparison with the K-S [24], we consider the case $\beta \equiv 0$ and define $\alpha = L^2/4\pi^2$. The drift-free K-V equation is then

$$u_t + 4u_{xxxx} + \alpha[u_{xx} + \tfrac{1}{2}(u_x)^2 + (uu_x)_x]$$

$$- \frac{\alpha}{4\pi} \int_0^{2\pi} (u_x)^2\, dx = 0 \quad , \quad 0 \leq x \leq 2\pi \quad , \tag{2.6}$$

$$u(x + 2\pi,t) = u(x,t) \quad , \quad u(x,0) = u_o(x) \quad .$$

We consider only initial values with zero mean. This implies that there is no cavitation and that the liquid follows the interfacial motion [16, 17].

The computational explorations outlined in [24] and here have spurred efforts to prove that the K-S and K-V equations are rigorously equivalent to a finite dynamical system. The approach (introduced in [12-15]) consists of constructing a finite dimensional Lifschitz inertial manifold Σ in the phase space of the PDE such that

(i) Σ is invariant and has compact support; that is, if $(S(t,\cdot))_{t\geq 0}$ is the nonlinear semigroup associated with the initial value problem for the equations, then $S(t,\Sigma)$ is contained in Σ for all $t \geqq 0$.

(ii) All solutions converge exponentially to Σ. In particular, the universal attractor, X, is included in Σ and the dissipative PDE reduces on Σ to a finite dynamical system (called an inertial ODE). More details of this structure can be found in the paper by C. Foias, in these proceedings.

The existence of such an inertial manifold has been demonstrated [6, 14, 15, 29] for the K-S equation with Neumann boundary condition. In this case the dimension of Σ is less than $c\alpha^{1.75}$, where c is a constant independent of α. This can be compared with an upper bound for the fractal dimension of the universal attractor X [28], $d_f(X) \leq c\alpha^{0.75}$). It has been proven that all solutions of (1.1) with Neumann boundary conditions converge exponentially, as $O(\exp\text{-}c\alpha^5 t)$, to the inertial manifold Σ. Thus when $\alpha > 1$, the solution is practically on the inertial manifold almost immediately. This is the essence of the argument that the K-S equation is a paradigm of a PDE equivalent to a finite dynamical system. Similar results hold for the K-V equation [Nicolaenko, unpublished] and for a model of 2-D weak turbulence in shear flows [8].

The bifurcation catalogues and diagrams (Figs. 1 and 2), were created by systematically scanning large intervals in the bifurcation parameter α. The diagrams for the K-S equation (Figs. 1b and 2b) are described elsewhere in detail [24]. While keeping α fixed, we searched for different attractors by varying the initial conditions. We then tracked the domains of stability of each attractor with respect to the bifurcation parameter by varying α and reinitializing $v(x,0)$ to the final solution from the previous run with a different α. Many problems were recalculated several times with different grid resolutions and time truncation error criteria to ensure that the numerical solutions were converged within an acceptable accuracy.

A remarkable feature of both the K-S and K-V equations is the alternating sequence of intervals in α that contain either laminar behavior, where a fixed point is ultimately attracting, or persistent oscillatory and/or chaotic behavior. Let $I_j = [\alpha_j, \alpha_{j+1}]$ be the jth interval. If α_1 is the point where the first Hopf bifurcation occurs, then $I_0 = [0, \alpha_1]$. A static pitchfork-like steady-state bifurcation occurs at $\alpha = 4 < \alpha_1$ for the K-V equation (Fig. 1a). For j <u>even</u>, I_j is characterized by the ultimate decay to a globally attracting fixed point $\tilde{u}_q(x)$, $q = (j/2) + 1$, $j \geq 2$. These fixed points have most of their energy concentrated in the q-th mode. The higher harmonics appear with exponentially decreasing energy and the fixed point has a lacunary Fourier expansion:

$$\tilde{u}_q(x) = a_{1q} \cos qx + \varepsilon\, a_{2q} \cos 2qx + \varepsilon^2 a_{3q} \cos 3qx + \ldots + \varepsilon^{n-1} a_{nq} \cos nqx + \ldots , \qquad (2.7)$$

where $q = j/2+1$. Numerically, we have found that a_{1q} is $O(1)$ and $\varepsilon \cong 10^{-1}$. We call these sinks associated with I_j, j even, <u>cellular states</u>. When the Fourier expansion (2.7) of a

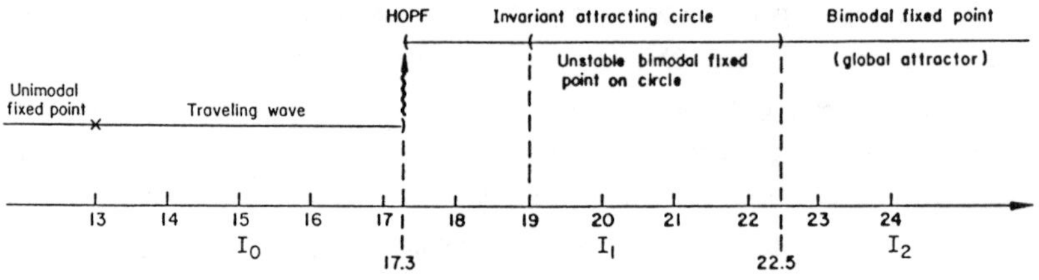

Fig. 1a. Stable solution of the K-S equation.

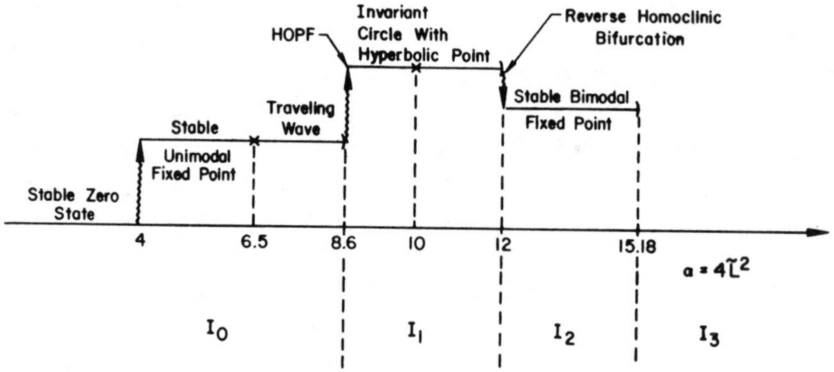

Fig. 1b. Stable solutions of the K-V equation.

Fig. 1. The stable solution manifolds for the K-S (Fig. 1a) and K-V (Fig. 1b) equation have a simple structure when α is small.

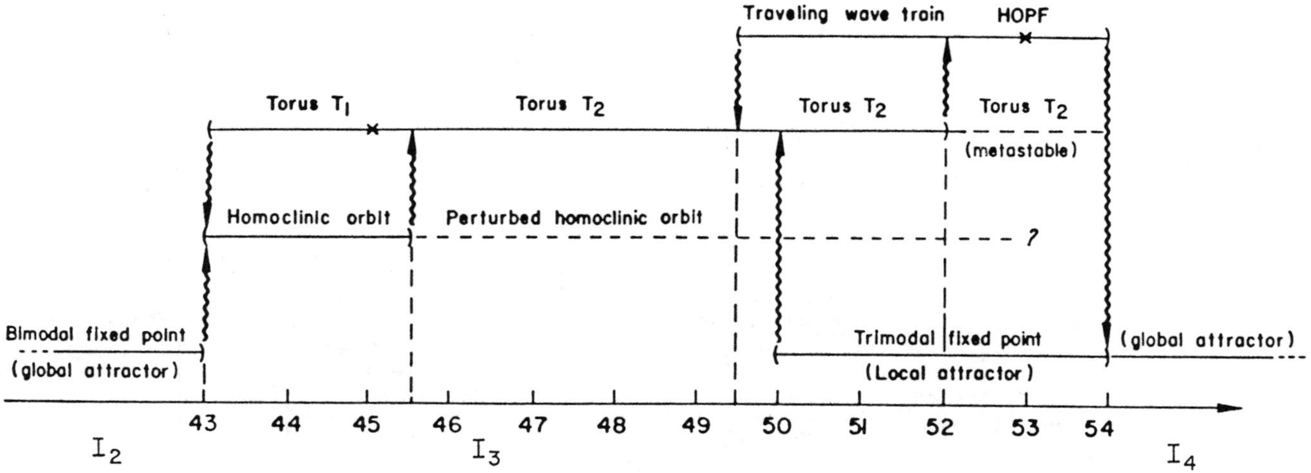

Fig. 2a. As the bifurcation parameter increases, the stable solution manifolds of the K-S equation become more complex.

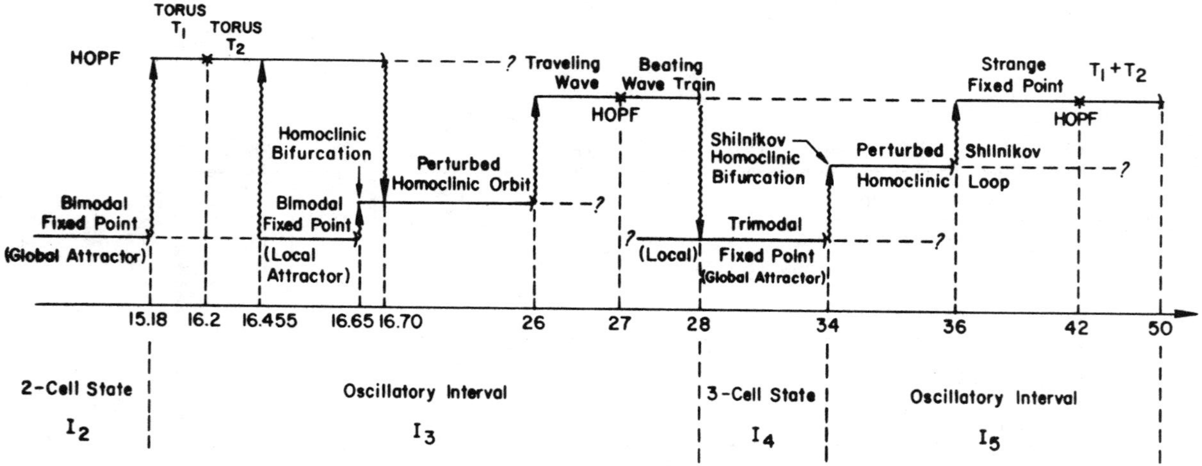

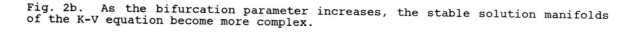

Fig. 2b. As the bifurcation parameter increases, the stable solution manifolds of the K-V equation become more complex.

cellular state is dominated by the q-th mode, we call it a q-modal cellular state.

The relaminarization intervals I_j, j even, are consistent with experiments at small and moderate Reynolds numbers [32]. Moreover, as j and α increase, the ultimate decay follows long periods of transient chaos. Transient chaos is observed in both the K-S and K-V equations beginning at the interval I_4, provided enough modes are excited in the initial data. Moreover, as α increases, the mean lifetime of transient chaos increases exponentially in $\tilde{L}$. When the fractal dimension of the universal attractor, X, for the flow is large, $\dim_f(X) \geqq 10$, this growth makes transient chaotic intervals undistinguishable, in practice, from chaotic intervals in the strongly chaotic regimes.

When j is odd, the intervals I_j have persistent oscillatory and/or chaotic behavior. For moderate values of α (up to about I_7), the quasi-periodic and/or chaotic behavior reflects a competition between the previous j+1/2 cellular state and the j+3/2 cellular state. This competition creates a complex interplay between temporal chaos and spatial coherence. In some sense, the (low-dimensional) temporal chaos in I_j corresponds to adjustment from one (low-dimensional) space pattern to the next one.

III. THE K-V EQUATION BIFURCATION INTERVALS

In this section, we describe the behavior of the solutions to the K-V equation for parameter values in the intervals I_j, $0 \leq j \leq 5$, $0 \leq \alpha \leq 50$ $(0 \leq \tilde{L} \leq 3.53)$. Within these intervals we see canonical vector field bifurcations leading to quasi-periodic motion and chaos. By systematically varying the initial conditions and α, we have constructed the preliminary bifurcation catalogue of the attractors to Eq. (2.6) shown in Figs. 1b and 2b.

These windows (intervals I_j) are much narrower than the corresponding ones for the K-S equation (Figs. 1a and 2a). Indeed, at α = 50, we are still in the quasi-periodic window I_3 for the K-S equation, but K-V has reached the end of the oscillatory window I_5. The term $(u_x)^2$ controls the blowup, but the nonlinearity $(uu_x)_x$ in the K-V equation accelerates the transition to chaos. Within each window, the bifurcation sequences for the K-S and K-V equations are generically similar: homoclinic loops, perturbed transverse homoclinic orbits and low-dimensional tori which eventually break up. In addition, here is a wealth of reverse bifurcations attractors that alternatively destabilize and restabilize again at some larger α.

In the discussion below, the "energy" is the integral of $(u_x)^2$ and the "energy in mode k" is the modulus of the kth Fourier coefficient.

The first bifurcation in the K-V equation is a classical pitchfork bifurcation at $\alpha = 4$ ($\tilde{L} = 1$) into a unimodal cellular state. At $\alpha = 6.50$, it bifurcates into a traveling wave (a similar phenomenon occurs in K-S at $\alpha = 13.0$). At $\alpha = 8.60$, the traveling unimodal wave undergoes Hopf bifurcation into an invariant circle. Surprisingly, this first sequence does not lead to further bifurcations and transition to chaos. The invariant circle remains attracting, yet a hyperbolic bimodal cellular point (repelling) is evident near $\alpha \cong 10$. At $\alpha = 12$ it undergoes a reverse homoclinic bifurcation [20] and the window I_1, $8.60 \leqq \alpha \leqq 12$, ends. In the interval I_2, $12 < \alpha < 15.18$, the bimodal fixed point, $\tilde{u}_2(x,\alpha)$, is globally attracting.

The second oscillatory window I_3 begins at $\alpha = 15.18$ where $\tilde{u}_2$ undergoes a nonclassical Hopf bifurcation that breaks the symmetry [18]. That is, the bifurcated circle breaks the group invariance $u(x,t) \to u(-x,t)$ of K-V. Tracking this torus T_1 (circle) by numerical continuation, we observed its bifurcation into T_2 (a two-dimensional torus) near $\alpha \cong 16.20$, with quasi-periodic dynamics (two incommensurate frequencies). This T_2 becomes metastable at $\alpha = 16.70$.

The attractor $T_1 + T_2$ is globally attracting for $15.18 \leq \alpha < 16.455$, but has only a limited basin of attraction for $16.455 \leq \alpha < 16.70$. At $\alpha = 16.455$, the bimodal cellular state $\tilde{u}_2$ undergoes its second reverse bifurcation and becomes a sink. It has a small basin of attraction on $16.455 < \alpha < 16.65$ where

$$\tilde{u}_2 = 5.9\cos 2x + 0.98\cos 4x + 0.09\cos 6x + \ldots \tag{3.1}$$

(at $\alpha = 16.455$) is stable with respect to small perturbations. The interval $16.455 \leqq \alpha \leqq 16.70$ is the first occurrence of coexisting attractors.

At $\alpha = 16.65$, a homoclinic loop appears with an infinite period orbit as $\tilde{u}_2$ undergoes its second (direct) homoclinic bifurcation [19]. As α increases, $\tilde{u}_2$ remains hyperbolic, with a nearly periodic, perturbed homoclinic orbit. The numerical example in Fig. 3 illustrates this for $\alpha = 22$, $u_0(x) = 6\cos 2x + \cos 4x$. The solution orbit spends a substantial time in a neighborhood of the saddle point

$$\tilde{u}_2(x) = 4.5\cos 2x + 1.15\cos 4x + 0.125\cos 6x + \ldots \quad . \tag{3.2}$$

The motion around the loop is triggered by a sensitive nonlinear exchange of energy between the odd and even modes. In Fig. 3b, the energy in Mode 1 bursts quickly from 10^{-7} to 10 as the solution traverses around the (perturbed) homoclinic loop. A corresponding dip is observed in the energy of Mode 2 (Fig. 3c).

These calculations were done with an error tolerance of 10^{-8} per unit timestep. When we relaxed the precision of the numerical integrations above 10^{-6}, the homoclinic loop disappeared and the solution locked into the numerically attracting bimodal fixed point. The high precision was necessary to trigger the energy feedback into the odd modes. Additional calculations with an error tolerance of 10^{-10} comfirmed that the homoclinic loop was not an artifact. This is one example (among many) that alerted us to the extreme sensitivity of the oscillatory solutions to the precision of the numerical method.

Similar dynamics associated with the stable and unstable manifolds of $\tilde{u}_2$ are observed until $\alpha \cong 26$, where a stable traveling wave train appears as a global attractor. This wave train is not related to either the previous bimodal point or to the trimodal cellular state, which is a global sink in I_4. It seems to be a special orbit on the metastable

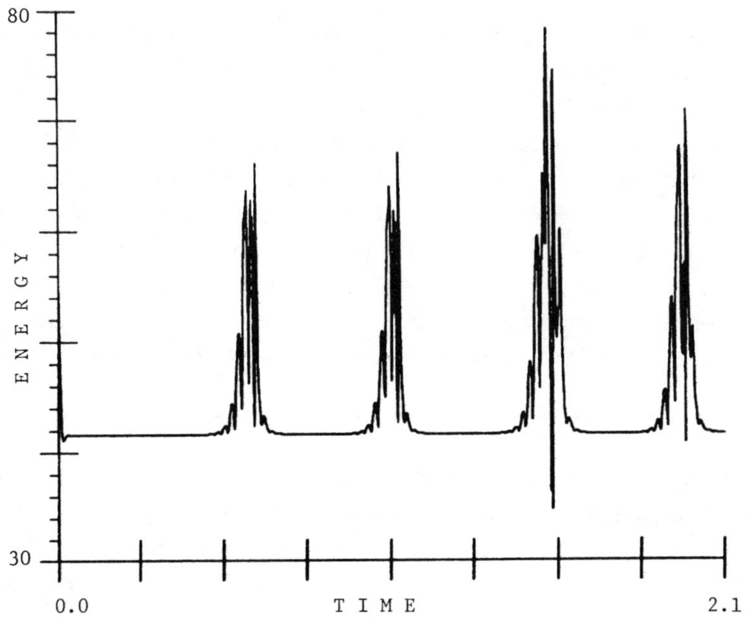

Fig. 3a. The energy of the solution has periodic bursts on the homoclinic loop (K-V, $\alpha = 27$).

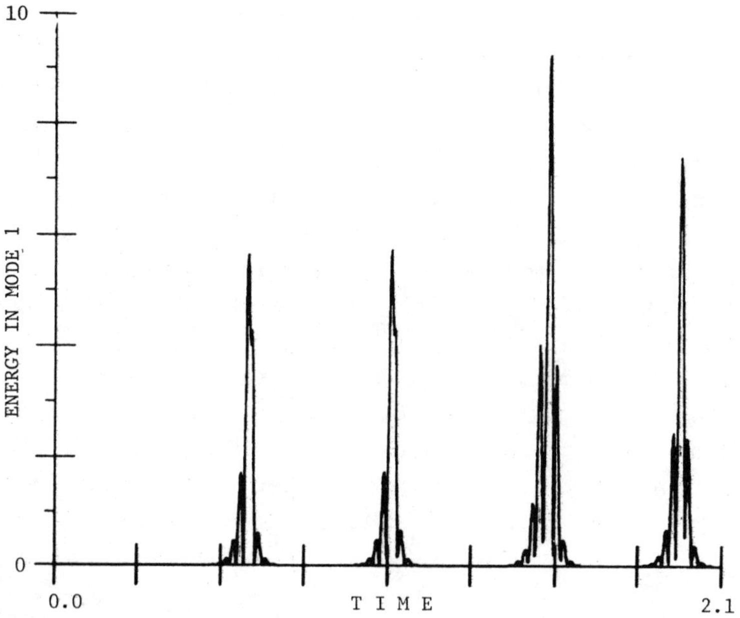

Fig. 3b. The energy in the first mode decays almost to zero near the bimodal saddle point.

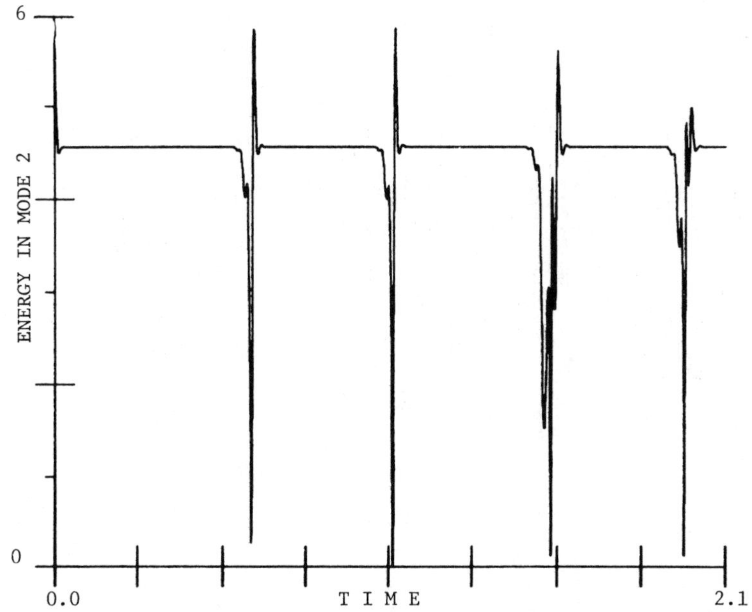

Fig. 3c. The energy in the second mode is nearly constant between the bursts. The saddle point has energy only in the even modes.

torus T_2. Its Fourier series expansion is not lacunary and a typical profile has two humps, a large one and a small one.

At $\alpha = 27$, the wave train undergoes Hopf bifurcation and becomes a strongly beating wave. The oscillations in the energy of the beating wave are evident in Fig. 4a ($\alpha = 27$, $u_0(x) = \cos x + \sin x + \cos 2x + \sin 2x + \cos 3x + \sin 3x + \cos 4x + \sin 4x$). The contour levels of the solution in Fig. 4b show the beating wave train drifting alternatively from left to right.

The window I_3 ends at $\alpha = 28$, where the trimodal cellular state

$$\tilde{u}_3 = 5.9 \cos 3x + 0.8 \cos 6x + \ldots \tag{3.3}$$

becomes a global sink. At $\alpha = 28$, initial conditions were imposed in which the first six cosine and sine modes were excited to a level of $O(1)$; the solution displayed persistent

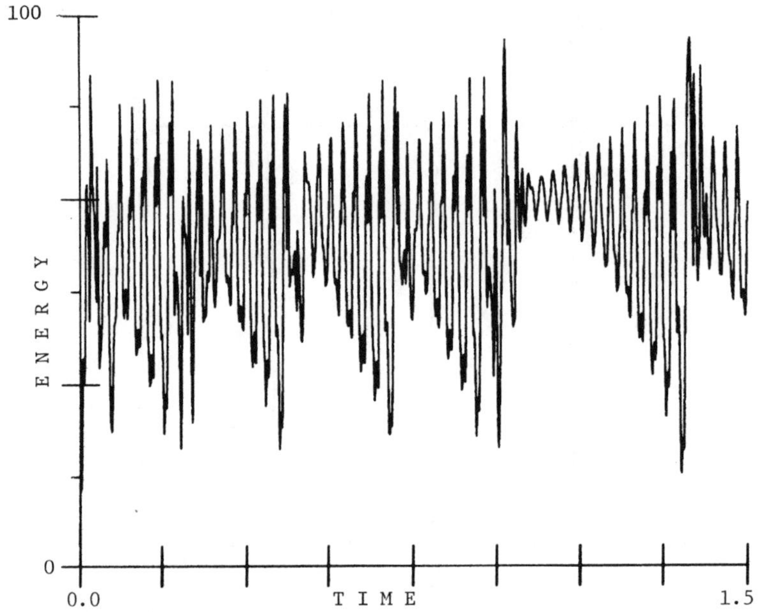

Fig. 4a. The traveling wave undergoes a Hopf bifurcation at $\alpha = 27$ (K-V) into a beating wave train.

Fig. 4b. The contour plot of the beating wave solution shows it drifting in alternate directions.

near-chaotic behavior before crashing abruptly into $\tilde{u}_3$. This type of transient behavior at the onset of an even-numbered interval heralds the transient chaos observed for large α's. The window I_4 spans $28 < \alpha < 34$.

These solutions linger in the background at higher α. The two-humped structure observed for $26 \leq \alpha \leq 27$ reoccurs in the midst of the next oscillatory interval I_5 (see Fig. 2b). At $\alpha = 36$, a steady, nondrifting, nonbeating fixed point reappears as a global attractor, with a similar two-humped profile (see Fig. 5). This strange fixed point has substantial energy in the first six modes and persists as a global sink until $\alpha = 42$, where it undergoes another Hopf bifurcation. The new Torus T_1 quickly evolves into T_2 and breaks down into near-chaotic dynamics. The transition into I_6 ($\tilde{u}_4$) occurs near $\alpha \cong 50$.

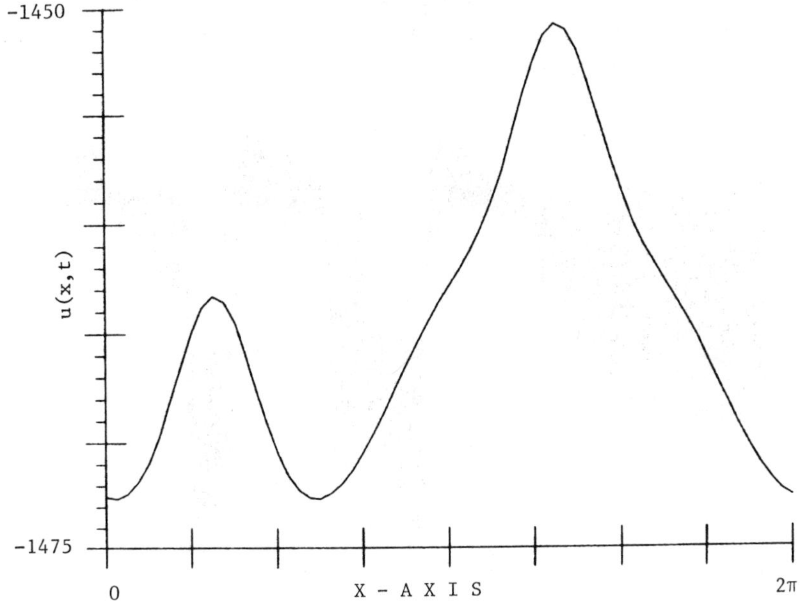

Fig. 5. This two-humped fixed point appears at $\alpha = 36$. Glimpses of it also appear in the traveling and beating waves at $26 < \alpha < 28$.

At $\alpha = 34$, the trimodal cellular state $\tilde{u}_3$ (a global sink in I_4) bifurcates. The bifurcation is neither of Hopf type, nor through a classical homoclinic loop. This is explored in Fig. 6 ($\alpha = 34$, $u_0 = 6\cos 3x + 0.8\cos 6x + 0.1\sin x$). At roughly periodic intervals, the orbit bursts away on the unstable manifold of $\tilde{u}_3$, intermittently puffs into chaos at a much lower energy level, and then spirals back around the hyperbolic point $\tilde{u}_3$. Fig. 6b confirms that the energy in the first mode is low during the small oscillations around the spiral hyperbolic point $\tilde{u}_3$ and is much higher during the chaotic bursts. The energy in the third mode, Fig. 6c, is the mirror image of Fig. 6b. It oscillates in a small neighborhood of 5.9, then bursts away from $\tilde{u}_3$ into a chaotic excursion. This behavior persists until $\alpha = 36$ and has many of the characteristics of a perturbed Shilnikov homoclinic loop [19] associated with a spiral hyperbolic point. To our knowledge, this is the first time it has been observed in a parabolic PDE. This behavior has also been observed and proved to exist for traveling wave solutions of the Fitz Hugh-Nagumo equation [25].

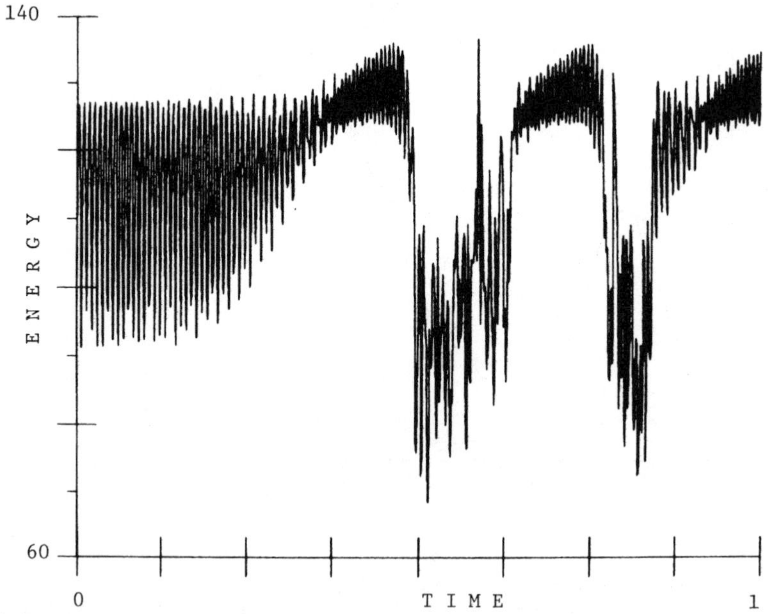

Fig. 6a. The energy has near-periodic bursts on the homoclinic loop and then spirals around the hyperbolic point (Shilnikov homoclinic loop, $\alpha = 34$, K-V).

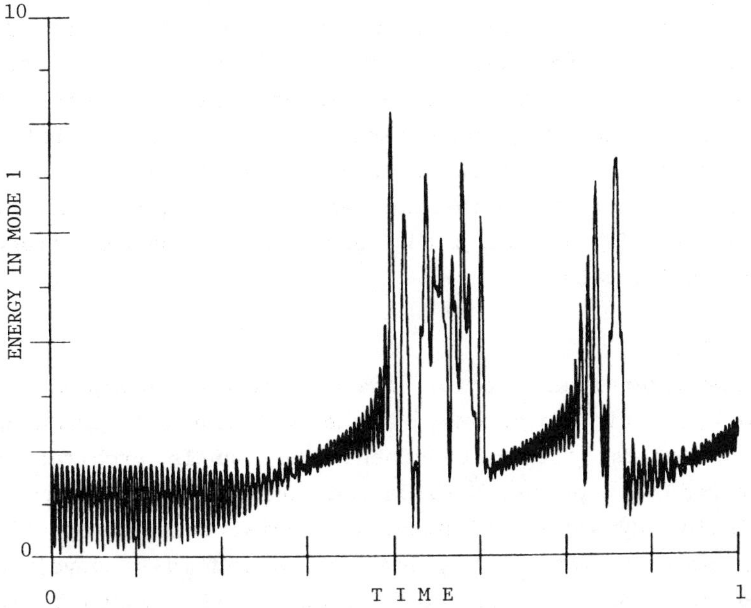

Fig. 6b. The energy in the first mode is low during the small oscillations near the trimodal hyperbolic point.

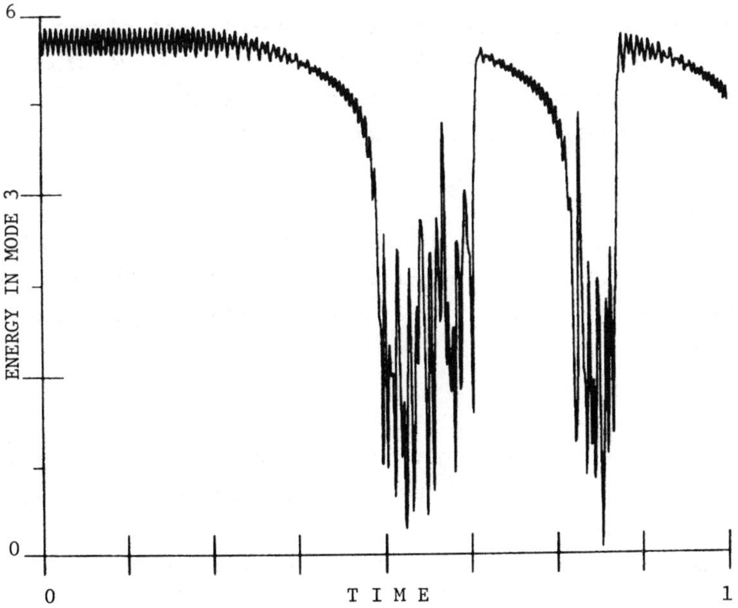

Fig. 6c. The energy in the third mode is high in the neighborhood of the spiral hyperbolic point, whose components contain energy only in the harmonics of three.

The bifurcations of the K-V equation, unraveled in this section, occur on low-dimensional inertial manifolds. Multiple forward and reverse bifurcations of several cellular fixed points are entangled in a web of tori, together with hyperbolic points. For these regimes, we conjecture that it may be possible to construct a simple reduced inertial normal form for the ODEs on the inertial manifold using the unstable manifolds of $\tilde{u}_2(x,\alpha)$, $\tilde{u}_3(x,\alpha)$, and the two-humped strange fixed point (Fig. 5).

IV. SUMMARY

A low-dimensional vector field skeleton underpins the sometimes chaotic solutions of the K-S and K-V turbulent interface models. This low-dimensional subtle architecture is mirrored by repeated bifurcations and intermittencies in the solution dynamics and plays a crucial role in bridging the gap between strong dynamical chaos and fully developed turbulence. Current analytic results [6] support the numerical evidence that this small exotic zoo of hyperbolic points and tori generates strong chaos in the K-S and K-V equations. We suspect that the dynamically relevant strange fixed points will be embedded in a Cantor-like structure in space (this has been proved for K-S, $\alpha = \infty$, by D. M. Michelson) and that spatial chaos will intermingle with temporal chaos at large values of the bifurcation parameter.

ACKNOWLEDGMENTS

The authors thank I. Kevrekidis for his helpful discussions and J. L. Castillo, P. L. Garcia-Ybarra, and M. G. Velarde for suggesting computational explorations of their model.

REFERENCES

1. P. Clavin, "Dynamic Behavior of Premixed Flame Fronts in Laminar and Turbulent Flows," Prog. Energy. Combust. Sci. 11 (1985), 1-59.
2. P. Constantine and C. Foias, Comm. Pure Appl. Math. 38 (1985), 1-27.

3. P. Constantine, C. Foias and R. Temam, Memoirs AMS 53 (1985) #314, vii+.67pp.
4. P. Constantine, C. Foias, O. P. Manley and R. Temam, C. R. Acad. Sci. Paris I, 297 (1983) 599-602.
5. P. Constantine, C. Foias, O. P. Manley and R. Temam, J. Fluid Mech. 150 (1985) 427-440.
6. P. Constantine, C. Foias, B. Nicolaenko and R. Temam, "Integral Manifolds and Inertial Manifolds for Dissipative PDE's (submitted).
7. J. D. Farmer, E. Jen, A. Brandstäter, J. Swift, H. L. Swinney, A. Wolff and J. P. Crutchfield, "Low Dimensional Chaos in a Hydrodynamic System," Phys. Rev. Lett. 51, 16 (1983) 1442-1445. See also J. P. Gollub and H. L. Swinney, Phys. Rev. Lett. 35 (1975) 927.
8. C. Foias, B. Nicolaenko and R. Temam, "Asymptotic Study of an Equation of G. I. Sivashinksy for Two Dimensional Turbulence of the Kolmogorov Flow," to appear, Proc. Paris Acad. Sci.
9. C. Foias and R. Temam, C. R. Acad. Sci. Paris, I, 295 (1982) 239-241.
10. C. Foias and R. Temam, C. R. Acad. Sci. Paris, I, 295 (1982) 523-525.
11. C. Foias and R. Temam, Mathematics of Computation 43 (1984) 117-133.
12. C. Foias, G. R. Sell and R. Temam, C. R. Acad. Sci. Paris, I, 301 (1985) 139-141.
13. C. Foias, G. R. Sell and R. Temam, "Inertial Manifolds for Dissipative PDE's (submitted).
14. C. Foias, B. Nicolaenko, G. R. Sell and R. Temam, C. R. Acad. Sci. Paris, I, 301 (1985) 285-288.
15. C. Foias, B. Nicolaenko, G. R. Sell and R. Temam, "Inertial Manifolds and an Estimate of Their Dimension for the Kuramoto-Sivashinsky Equation (submitted).
16. P. L. Garcia-Ybarra and M. G. Velarde, "Oscillatory Marangoni-Benard Interfacial Instability and Capillary-Gravity Waves in Single and Two-Component Liquid Layers with or without Soret Thermal Diffusion," to appear.
17. P. L. Garcia-Ybarra, J. L. Castillo and M. G. Velarde, "Benard-Marangoni Convection with a Deformable Interface and Poorly Conducting Boundaries," to appear.

18. M. Golubitsky and D. G. Schaeffer, "Singularities and Groups in Bifurcation Theory," Springer-Verlag, New York 1985).

19. J. Guckenheimer and P. H. Holmes, Nonlinear Oscillations, Dynamical Systems and Bifurcation of Vector Fields, Springer-Verlag (1985).

20. J. Guckenheimer, "Strange Attractors in Fluids: Another View," Annual Review of Fluid Mechanics (1986).

21. J. M. Hyman, "MOLID: A General Purpose Subroutine Package for the Numerical Solution of Partial Differential Equations," Los Alamos Scientific Laboratory Manual, LA-7595-M (1978).

22. J. M. Hyman, "Numerical Methods for Nonlinear Differential Equations," Nonlinear Problems: Present and Future, A. R. Bishop, D. K. Campbell and B. Nicolaenko, Eds., North-Holland Publ. Co., (1982) 91-107.

23. J. M. Hyman, "Numerical Methods Based on Analytic Approximations" (in preparation).

24. J. M. Hyman and B. Nicolaenko, "The Kuramoto-Sivashinsky Equation: a Bridge Between PDE's and Dynamical Systems," to appear in Physica D (1986).

25. Guan-Hsong Hsu, Ph.D. Thesis, 1985, Courant Institute of Math. Sciences, New York University.

26. B. Nicolaenko and B. Scheurer, "Remarks on the Kuramoto-Sivashinsky Equation," Physica 12D (1984) 331-395.

27. B. Nicolaenko, B. Scheurer, and R. Temam, "Quelques proprietes des attracteurs pourl'equation de Kuramoto-Sivashinsky," C. R. Acad Sci. Paris 298 (1984) 23-25.

28. B. Nicolaenko, B. Scheurer, and R. Temam, "Attractors for the Kuramoto-Sivashinsky Equations," Physica 16D (1985) 155-183.

29. B. Nicolaenko, B. Scheurer, and R. Temam, "Attractors for the Kuramoto-Sivashinsky Equations," AMS-SIAM Lectures in Applied Mathematics 23, 2 (1986) 149-170.

30. B. Nicolaenko, B. Scheurer, and R. Temam, "Attractors for Classes of Nonlinear Evolution of Partial Differential Equations," in preparation.

31. G. I. Sivashinsky and A. Novick-Cohen, "Interfacial Instabilities in Dilute Binary Mixtures Change of Phase, to appear in Physica D.
32. K. Shreenivasan, "Transition and Turbulence in Fluid Flows and Low-Dimensional Chaos," Frontiers in Fluid Mechanics, S. H. Davis and J. L. Lumley, Eds., Springer-Verlag (1985) 41-67.
33. R. Temam, Navier Stokes Equations and Nonlinear Functional Analysis, SIAM, Philadelphia, 1983.
34. R. Temam, "Infinite Dimensional Dynamical Systems of Fluid Mechanics," AMS-Summer Res. Institute "Nonlinear Funct. Anal. Appl." (Berkeley, 1983).
35. M. G. Velarde and C. Normand, Sci. Amer. 243 (1980) 92.
36. M. G. Velarde and J. L. Castillo, "Convective Transport and Instability Phenomena, J. Zierep and H. Oertel, Jr., Eds., Braun-Verlag, Karlsruhe (1982).

The authors were supported by the US Department of Energy under contract W-7405-ENG-36 and the Office of Scientific Computing under contract KC-07-01-01-0.

Center for Nonlinear Studies
Theoretical Division, MS B284
Los Alamos National Laboratory
Los Alamos, NM 87545

Einstein Geometry and Hyperbolic Equations

S. Klainerman

One of the main goals in the theory of differential equations of mathematical physics is to describe the behavior of general solutions to the Cauchy problem. In a general set-up (in the absence of boundary conditions) this amounts to finding solutions $u = (u^1,\dots,u^N)$ depending on $x = (x^0,x^1,\dots,x^n) \in R^{n+1}$ which satisfy a system of N independent equations

$$G(x,J^m u(x)) = 0 \tag{G}$$

and whose partial derivatives of order $\leq m-1$ are prescribed on a hypersurface $\underline{H}^n \subset \mathbb{R}^{n+1}$. Here $J^m u = (u,\nabla u,\dots,\nabla^m u)$ denotes all partial derivatives of u of order less than or equal to m and $G = (G^1,\dots,G^N)$ are smooth, given functions of x, $J^m u$. The basic physical systems are invariant under translations, i.e. G does not depend explicitly on x, and linear in the highest derivatives $\nabla^m u$. Moreover m = 1 or 2.

The first m-1 derivative of u on $\underline{H}^n$ are uniquely determined by the first m-1 normal derivatives i.e.

$$T^j u|_{\underline{H}^n} = \psi^j , \qquad j = 0,1,\dots,m-1 , \tag{C.D.}$$

with T the derivative in the direction of the unit normal to $\underline{H}^n$. They are called the Cauchy data for u, i.e. $\text{Cauchy}_{\underline{H}^n}(u) = (\psi^0, \psi^1,\dots, \psi^{m-1})$. The derivatives of order $\geq m$ can be

ISBN 0-12-195255-X

formally determined, in a unique fashion, from the Cauchy data provided that $\underline{H}^n$ is *noncharacteristic*. This amounts to a simple algebraic condition on the derivatives of F relative to $\nabla^m u$ on the given Cauchy data. Moreover this formal procedure can be made rigorous provided that we restrict ourselves to real analytic solutions and real analytic F, $\underline{H}^n$ and Cauchy data. The result is the famous theorem of Cauchy-Kowalewski which allows one to find unique real analytic solutions in the neighborhood of a point in $\underline{H}^n$, or a whole strip around it (Fig. 1).

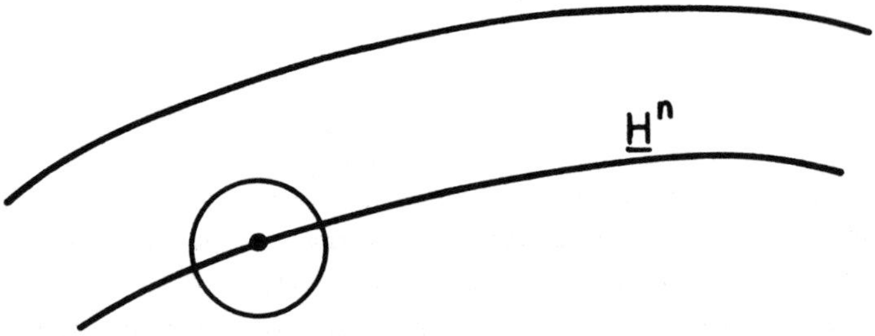

Figure 1

Now the class of real analytic solutions is too limited to describe the evolution of physical systems. Indeed a real analytic function is uniquely determined by the values in any small neighborhood of a point. It is therefore impossible to describe the propagation of local perturbations of the Cauchy data by using real analytic solutions. Moreover Einstein's Principle of Relativity postulates that physical signals cannot propagate with velocity larger than the speed of light. Consequently, acceptable equations, modeling physical systems, have to satisfy the property of *finite speed of propagation*. Roughly, we say that this property is satisfied if, given a solution u and a relatively compact set D in $\underline{H}^n$ one can find a cone C(D) (as in Fig. 2) such that if v is any other solution with Cauchy(v) = Cauchy(u) on $\underline{H}^n \setminus D$ then u = v in $\mathbb{R}^{n+1}\setminus C(D)$. $\underline{H}^n$ is said to be spacelike relative to u.

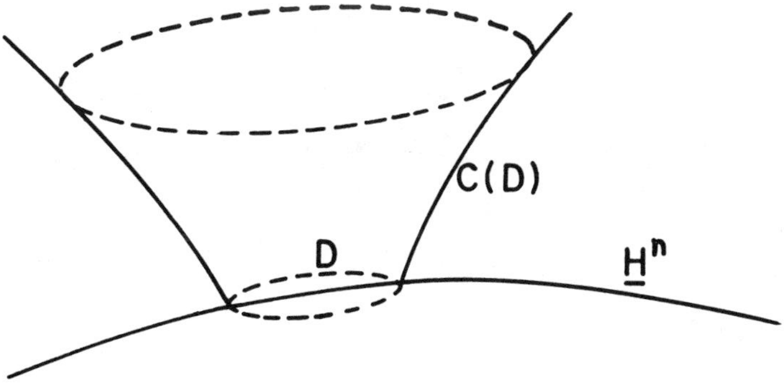

Figure 2

In particular, the property of finite "speed of propagation" requires that the Cauchy Problem be solved in function spaces with good local properties like C^∞, C^k and so on. If this can be achieved we say that the Cauchy Problem is "well posed". This calls for much more stringent algebraic assumptions on the system. In different formulations they are called hyperbolicity assumptions. For nonlinear equations a good notion of hyperbolicity leads to a local existence and uniqueness theorem for C^∞ or C^k spaces of Cauchy data. One accomplishes this in a neighborhood of a given solution $\bar{u}$. The main step in the proof of such a theorem is

Step 1. "A-priori L^2 (energy) estimates".

These estimates depend on a splitting of $\mathbb{R}^{n+1}$ (in a neighborhood of $\underline{H}^n$) between a time direction t transversal to $\underline{H}^n$ and spatial directions "parallel" to $\underline{H}^n$. More precisely given $\bar{u}$ and $\underline{H}^n$ noncharacterstic and space-like relative to $\bar{u}$ we can introduce a locally defined time function $t = t(x^0, x^1, \ldots, x^n)$ such that $\underline{H}^n$ is given by $t = 0$ and each level surface $\underline{H}^n_c = \{x \mid t(x)=c\}$ is noncharacteristic and space-like relative to all solutions u close to $\bar{u}$,

$$|J^{m-1}u - J^{m-1}\bar{u}| < \delta$$

with $\delta > 0$ sufficiently small, and for $0 \leq t \leq t_*$ (see Fig. 3).

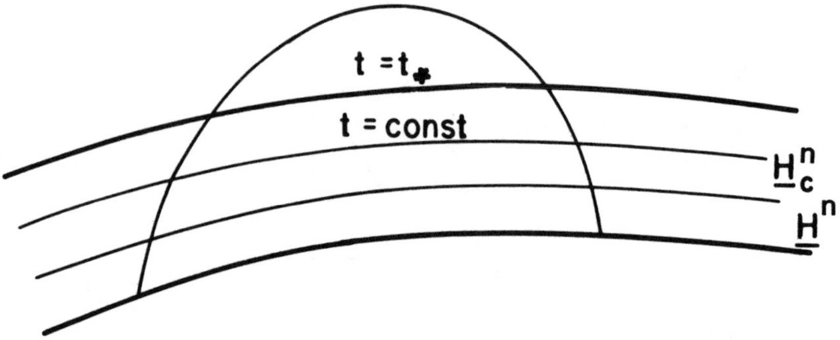

Figure 3

Given v, a function on $\mathbb{R}^{n+1}$, we define the L^2 norms

$$||v(t)||_s = \sum_{0\le k\le s} \int_{\underline{H}_t^n} |\nabla^k v(x)|\, d\sigma_t \tag{1}$$

with $d\sigma_t$ the volume element of $\underline{H}_t^n$. Also define the L^∞ norms,

$$|v(t)|_s = \sum_{0\le k\le s} \sup_{\underline{H}_t^n} |\nabla^k v(x)| \,. \tag{2}$$

By so-called energy estimates one establishes an a-priori inequality which has, essentially, the following form

$$||v(t)||_s \le ||v(0)||_s \exp\left(C_s \int_0^t |v(\tau)|_m\right) \tag{3}$$

for all $0 \le t \le t_*$, with $v = u - \bar{u}$ and the constant C depending on $\sup_{0\le t\le t_*} |v(t)|_{m-1}$ and certain bounds on $\bar{u}$. *

---

*Estimates of this type were discussed in [1] (for second order equations, or symmetric hyperbolic systems) and are based on the sharp product and chain rule inequalities in Sobolev norms (see [2], [3], [4], [5]).

Step 2. "Sobolev Inequalities".

The main idea in this step is to try to control the uniform norm $|v(t)|_m$ in terms of the L^2-norms $||v(t)||_s$. One has, by the classical Local Sobolev Inequalities

$$|v(t)|_m \leq c_m ||v(t)||_{m+\underline{n/2}}$$

with $\underline{n/2}$ the smallest integer > n/2. Therefore, inserting this in (3),

$$||v(t)||_s \leq ||v(0)||_s \exp(c_s \int_0^t ||v(\tau)||_{m+\underline{n/2}} d\tau).$$

Finally, set $s_0 = m+\underline{n/2}$ and use a contraction argument relative to the norm $\sup_{0 \leq t \leq t_*} ||v(t)||_{s_0}$. This will prove the existence of a local solution u, for any Cauchy data sufficiently close to $\text{Cauchy}_{\underline{H}^n}(\bar{u})$ relative to the Sobolev norm $||\ \ ||_{s_0}$, provided that $t_* ||v(0)||_{s_0}$ is sufficiently small (see [2], [3], [4], [5]). The critical number s_0 can be made smaller than $m+\underline{n/2}$ provided that the system is nonlinear in a milder sense.

The concepts of hyperbolicity, finite propagation speed, space-like hypersurfaces and the local existence proof sketched above took many years and many mathematicians to complete. To go beyond them one needs to look much closer to the specific properties of physical systems. In particular most such systems have certain conserved quantities. The one which is very often used is a positive integral on space-like hypersurfaces, depending on (some of) the first m-1 derivatives of u, and which is called the (total) energy of the system. The positivity of the energy is usually obvious, however in the particular case of General Relativity it required a nontrivial beautiful theorem proved by Schoen-Yau [1] and Witten [2]. (The concept of energy itself is somewhat more elusive in General Relativity. To make sense of it one needs to introduce the equally elusive notion of an isolated system which will be discussed below.) In any case, the energy allows one to derive a global bound on L^2-norms of some of the first m-1 derivatives of u. When the system is only "mildly nonlinear" this can be used to prove a global existence theorem, in the whole space-time $\mathbb{R}^{n+1}$ and for any smooth Cauchy data. Such results were recently proved for the Yang-Mills equation by Eardley-Moncrief [8], or earlier for semilinear Klein-Gordon equations by Jörgens [9]. Yet for systems with strong nonlinearities the bound provided by the energy is not sufficient to draw any conclusions on the global behavior of solutions. This is the case of the systems describing such fundamental physical theories as Elasticity, Fluid Dynamics or General Relativity. Moreover these systems do not have global smooth solutions for arbitrary data. In fact most solutions, which start by being perfectly smooth on the initial hypersurface, will become singular after only a short lapse of time. The singularities can be so strong as to prohibit their continuation, as classical solutions, as is the case of <u>shock waves</u> in Fluid Dynamics, or any continuation at all, as for curvature singularities inside <u>black holes</u> in General Relativity. So the next fundamental question in the study of the Cauchy Problem is to understand when and how these singularities form, to describe the behavior of solutions near them and find a way, if possible, to continue the solutions beyond the singularities in a physically meaningful fashion. A simpler, related, question is to find under what circumstances singularities may be avoided altogether. Indeed all physical systems have trivial global solutions, corresponding to no action at all like the zero solution $\bar{u} = 0$. If the system G is autonomous and has $\bar{u} = 0$ as a solution we can rewrite it in the form

$$Lu = F(J^m u) \qquad \text{(F)}$$

with L a linear m^{th} order operator with constant coefficients and F a smooth function of $J^m u$, linear in $\nabla^m u$ and vanishing together with the first derivatives for $J^m u = 0$. For small Cauchy data,

$$\partial_t^j u = \varepsilon\psi^j , \qquad j = 0,1,\ldots,m-1 \qquad \text{(C.D.)}$$

the local existence theorem assures the existence of a smooth solution in a time interval $[0,t_*]$ of size $O(1/\varepsilon)$. To go beyond this one may want to compare the solutions to (F) with those of the linear problem

$$Lu = 0 \, . \qquad \text{(L)}$$

Now this approach has been quite successful in the study of boundary value problems for nonlinear elliptic equations where the linear part is manifestly dominant. However, for hyperbolic equations the behavior of (L) and (F) differ even locally (as one can see by comparing their corresponding characteristics). Yet certain global properties of (L) can be used for nonlinear equations. In particular, looking at the energy estimate (3), with $\bar{u} = 0$, we remark that if one could establish that $|u(\tau)|_m$ decays sufficiently rapidly as $\tau \to \infty$ then $||u(t)||_s$ is bounded uniformly in $t \geq 0$. This would allow us to extend the solutions for all time. Now it is true that most linear equations (L) of importance in mathematical physics have some uniform decay properties. This is a consequence of the spreading of physical signals in the whole space and can be quite easily verified by Fourier methods or, directly, from the form of the explicit solutions. In particular, when L is the usual wave operator, i.e. $L = -\Box = \partial_t^2 - \partial_1^2 - \ldots - \partial_n^2$, in $\mathbb{R}^{n+1}$ and $\Box u = 0$ one has the estimate

$$|u(t,x)| \leq C(1 + t)^{-(n-1)/2} \qquad (4)$$

uniformly in $x \in \mathbb{R}^n$, $t \geq 0$. The constant C depends on the L^1 norm of the first n/2 derivatives of the initial data (see [2], [10]). Such estimates have been used to derive global results for scalar nonlinear wave equations. I would like to mention in particular the early work of I. Segal [11] and W. Strauss [12] for semilinear cubic scalar equations (i.e. $F(u) = O(|u|^3)$ for small u) and by F. John ([13], [14]), J. Shatah [15] and S. Klainerman ([1], [2]) for general classes of scalar wave equations. However, the dependence of the constant C on the L^1-norm of the initial data has seriously restricted the use of (4) to treat quadratic nonlinearities. More recently, in [16], we have presented a method, based on the Lorentz and scale invariance of $\Box$, to derive the same rates of uniform decay in t directly from some generalized version of energy inequalities. In fact the idea was to modify the classical Sobolev inequality used in step 2 of the existence proof outlined below, by a global version of it with $\| \ \|_s$ replaced by a larger weighted Sobolev norm (see (15)) and multiplied by a decay factor of order $O((1+t)^{-(n-1)/2})$ as $t \to \infty$. These new Sobolev norms are bounded when applied to solutions of $\Box u = 0$, which allows us to derive the same rates of decay for u as in (4). Moreover they can be applied to nonlinear equations, $\Box u = F(u,\nabla u,\nabla^2 u)$, to derive estimates similar to those in (3). As a consequence one can follow the same steps as those of the local existence theorem to derive global existence, if $(n-1)/2 > 1$, or "almost global existence", if $(n-1)/2=1$, provided that the initial conditions are sufficiently small (see [16], [17], also [18] and the more recent results of [19], [20]).

The aim of this lecture is to discuss the relation between the geometry of the Minkowski space, generalized energy estimates and uniform rates of decay for two other examples of linear field equations in Minkowski space

(M) Maxwell equations

(Sp) Spin-2 field equations.

For completeness we will also consider

($\Box$) Scalar wave equation.

Our main interest is the study of (Sp) and its connection to an outstanding problem in general relativity which we state below. All the results presented in this lecture were derived in collaboration with D. Christodoulou [21].

We first recall a few basic concepts in Einstein Geometry.

A differentiable manifold M^{n+1} of dimension n+1 is called an Einstein manifold if it has a nondegenerate metric $< \ , \ >$ of signature $(-1,1,\ldots,1)$. In local coordinates $(x^0,x^1,\ldots,x^n)$,

$$ds^2 = g_{\mu\nu}\, dx^\mu\, dx^\nu, \text{ with } g_{\mu\nu} = <\frac{\partial}{\partial x^\mu}, \frac{\partial}{\partial x^\nu}>.$$

Denote by $\mathfrak{X}(M^{n+1})$ the set of all vector fields X on M^{n+1},

$$X = X^{\mu}(x) \frac{\partial}{\partial x^{\mu}} ,$$

with,

$$\langle X,Y\rangle = g_{\mu\nu} X^{\mu}Y^{\nu} = X^{\mu}Y_{\mu} .$$

We say that X is time-like, null or space-like depending on whether $\langle X,X\rangle$ is negative, zero or positive. A hypersurface $\underline{H}^n$ is called space-like if its normal is time-like at every point of $\underline{H}^n$. The metric induced by $\langle\ ,\ \rangle$ on a space-like hypersurface is Riemannian.

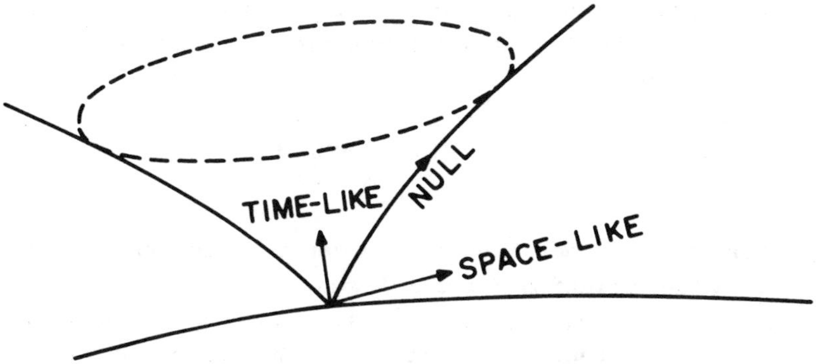

Figure 4

A frame $e_0,e_1,\ldots,e_n$ is called orthonormal if $\langle e_{\mu},e_{\nu}\rangle = \eta_{\mu\nu}$ for $\mu,\nu = 0,1,\ldots,n$. We will also consider null frames i.e. frames $e_1,\ldots,e_{n-1}, e_n, e_{n+1}$ with e_n, e_{n+1} null, $\langle e_n,e_{n+1}\rangle = -1$ and $e_1,\ldots,e_{n-1}$ orthonormal vectors on the space complementary to e_n and e_{n+1}. Relative to an arbitrary frame (e_{μ}), $\mu = 0,\ldots,n$, the components of a tensor W are $W_{\mu_1 \ldots \mu_p}$ and we raise and lower indices with the help of $g_{\mu\nu} = \langle e_{\mu},e_{\nu}\rangle$ and $g^{\mu\nu}$, the matrix inverse of $g_{\mu\nu}$. The Lie derivative of a tensor W relative to a vector field $X \in \mathfrak{X}(M)$ will be denoted by $\mathcal{L}_X W$. As in Riemannian geometry, there exists a unique affine connection compatible with the metric i.e. for $X,Y,Z \in \mathfrak{X}(M)$

$$\nabla_X Y - \nabla_Y X = [X,Y]$$

and

$$\nabla_X\langle Y,Z\rangle = \langle\nabla_X Y,Z\rangle + \langle Y,\nabla_X Z\rangle .$$

Relative to a frame $(e_\mu)_{\mu=0,\dots,n}$ the components of the covariant derivative ∇W of a tensor W are $W_{\mu_1 \dots \mu_p;\mu_{p+1}}$. Again as in Riemannian Geometry, the Riemann tensor $R_{\alpha\beta\gamma\delta}$ is obtained by,

$$(\nabla_X \nabla_Y - \nabla_Y \nabla_Y) Z = R(X,Y)Z + \nabla_{[X,Y]} Z$$

for any $X,Y,Z \in \mathfrak{X}(M)$.

The basic geometric operators for tensors, like the D'Alembertian $\Box_g$ and Divergence Div_g, are obtained from covariant differentiation and contractions e.g. for vectors $V \in \mathfrak{X}(M)$, or symmetric and antisymmetric 2-tensors W,

$$\mathrm{Div}\, V = g^{\alpha\beta} V_{\alpha;\beta}$$
$$(\mathrm{Div}\, W)_\alpha = g^{\beta\gamma} V_{\alpha\beta;\gamma} \quad *$$

For a scalar u we define the D'Alembertian

$$\Box_g u = g^{\alpha\beta} u_{;\alpha\beta} .$$

The other basic operation is the generalized curl, or the exterior differentiation d for p-forms.

The field equations are systems of differential equations for tensors, obtained by using covariant differentiation, exterior differentiation and contractions. As far as they lead to determined systems of equations, the corresponding Cauchy Problems are, typically, well posed on any"space-like" hypersurface. So in Einstein spaces the differential equations which are tied to the geometry are automatically hyperbolic.

The simplest example is that of the scalar wave equation

$$\Box_g u = 0 . \qquad (\Box)$$

The Cauchy problem,

$$u = \psi_{(0)}, \quad Tu = \psi_{(1)} \quad \text{on } \underline{H}^n$$

is well posed for any space-like hypersurface $\underline{H}^n$.

---

* We will use the notation div for the induced Riemannian metric on a space-like hypersurface.

The Maxwell equations are expressed, in a four-dimensional space-time M^4, with the help of an antisymmetric tensor $F_{\mu\nu}$ called the electromagnetic tensor. Any such F can be decomposed into two vector fields orthogonal to a given direction $X \in \mathfrak{X}(M)$ by forming the contractions

$$(i_X F)_\mu = F_{\mu\nu} X^\nu \ , \qquad (i^*_X F)_\mu = {}^*F_{\mu\nu} X^\nu \ ,$$

where ${}^*F$ is the Hodge dual to F, i.e. ${}^*F_{\alpha\beta} = \frac{1}{2}\varepsilon_{\alpha\beta\mu\nu} F^{\mu\nu}$ and $\varepsilon_{\alpha\beta\gamma\delta}$ the volume form on M^4. They determine the full tensor F whenever $\langle X,X\rangle \neq 0$, and are called the electric and magnetic components of F.

The four dimensional form of the Maxwell equations is

$$F_{\alpha\beta;\gamma} + F_{\beta\gamma;\alpha} + F_{\gamma\alpha;\beta} = 0 \qquad (M_i)$$

$$F_{\alpha\beta}{}^{;\beta} = 0 \ . \qquad (M_{ii})$$

The Cauchy problem is given by

$$i_T F = E_{(0)} \ , \quad i^*_T F = H_{(0)}$$

on any space-like hypersurface $\underline{H}^3$ with unit normal T and $E_{(0)}$, $H_{(0)}$ given vectors in $\underline{H}^3$ satisfying the constraints

$$\text{div } E_{(0)} = \text{div } H_{(0)} = 0 \quad \text{on} \quad \underline{H}^3 \ .$$

Before defining the Spin-2 equations we will give below a short motivation of their interest for us. They are connected to the Bianchi identities satisfied by the Riemann curvature tensor of Einstein-Vacuum space-times which we describe below.

Einstein Equations in Vacuum.

The Einstein field equations were proposed by A. Einstein 1916 as a unified theory of space-time and gravitation. The four manifold M^4 is itself an unknown, one has to find an Einstein metric $g_{\mu\nu}$ such that

$$R_{\mu\nu} - \frac{1}{2} g_{\mu\nu} R = Q_{\mu\nu}$$

where $R_{\mu\nu}$ is the Ricci tensor of the metric, $R_{\mu\nu} = R^{\alpha}_{\mu\alpha\nu}$, $R = g^{\mu\nu} R_{\mu\lambda}$ the scalar curvature, and $Q_{\mu\nu}$ is the energy momentum tensor of a matter field (e.g. the Maxwell equations). Contracting twice the Bianchi identities

$$\nabla_{[\varepsilon}R_{\alpha\beta]\gamma\delta} := \nabla_{\varepsilon}R_{\alpha\beta\gamma\delta} + \nabla_{\alpha}R_{\beta\varepsilon\gamma\delta} + \nabla_{\beta}R_{\varepsilon\alpha\gamma\delta} = 0$$

we derive $\nabla^{\nu}(R_{\mu\nu} - \frac{1}{2} g_{\mu\nu}R) = 0$ which are equivalent with the divergence equations of the matter field.

$$\text{Div } Q = 0 .$$

In the simplest situation (i.e. $Q_{\mu\nu} = 0$), of physical vacuum, the Einstein equations take the form

$$R_{\mu\lambda} = 0 \tag{E.V.}$$

called the Einstein Vacuum Equations. Written explicitly, in an arbitrary system of coordinates, they lead to a degenerate system of equations. However, since the equations are invariant under any diffeomorphisms of M^4 it suffices to find coordinate conditions called also gauge conditions, which lead to a well posed Cauchy problem. This was achieved by Y. Choquet-Bruhat in the harmonic gauge [22], yet, as she pointed out later [23], the harmonic gauge is unstable in the large. This problem of finding a globally stable, "well posed", gauge condition (whether it exists at all) is the first major difficulty one has to overcome when trying to construct global space-times.

One particular solution of (E.V.) is the Minkowski space $\mathbb{R}^{3+1}$ i.e. $\mathbb{R}^4$ with a given canonical coordinate system (x^0,x^1,x^2,x^3) and metric $< , >$ such that

$$<\partial_{\mu}, \partial_{\nu}> = \eta_{\mu\nu} = \text{Diag}(-1,1,\ldots,1) .$$

So we would like to ask whether Cauchy developments, i.e. solutions to Cauchy data on a given 3-manifold, which are small in an appropriate sense, lead to global, complete, smooth, solutions to the Einstein vacuum equations. We expect these solutions to be close, in a certain sense, to the Minkowski space. More precisely they should be "asymptotically flat space-times" as they are known in the Physics literature. The study of such space-times, as stressed by Geroch [24], is the study of isolated systems. They are of great interest in General Relativity since "it is only through a suitable notion of an isolated system that one acquires an ability at all to deal individually with various subsystems in the universe; in particular to assign to subsystems such physical attributes as mass, angular momentum, character of emitted radiation, etc." Yet it is not at all obvious how to even formulate the notion that a space time is close to the Minkowski space. (This task is comparatively easier in Riemannian Geometry due to the directional uniformity of the Euclidean space). However a global notion of an asymptotically flat manifold has been developed in the last 20 years (see [25] for a survey) beginning with the work of Bondi [26], [27] (see also Sacks [28]) who introduced the idea of analyzing solutions of the field equations along null, or characteristic, surfaces of a space-time. The present state of understanding was set by Penrose ([29], [30]) who formalized the idea of asymptotic flatness by adding a boundary, attached by the process of a conformal compactification. In fact, Penrose defined a space-time to be asymptotically flat if this boundary could be attached in a regular fashion.

With all the appealing features of the idea of regular, conformal compactification it is not at all clear that there are any, nontrivial solutions (involving gravitational radiation) satisfying the Penrose requirements. We believe that a more complete understanding of asymptotically flat space-times can only be accomplished by constructing them from given initial data. Our aim is to achieve that for sufficiently small data.

To make the problem more precise one has to introduce the notion of "space-like asymptotically flat" manifolds. We say that a Riemannian 3-manifold $\underline{H}^3$, diffeomorphic to $\mathbb{R}^3$, is asymptotically flat if there exists a coordinate system (x^1,x^2,x^3) globally defined, with the possible exception of a relatively compact set, such that the metric has the asymptotic properties

$$g_{ij} = \left(1 + \frac{M}{2r}\right) \delta_{ij} + o(r^{-1}) \qquad (5)$$

as $r = [\sum_{i=1}^{3} (x^i)^2]^{1/2} \to \infty$, and M a constant, called the mass of $\underline{H}^3$.

So we will try to construct global (E-V) 4-manifolds which are asymptotically flat on any space-like hypersurface. We will call them space-like asymptotically flat or simply asymptotically flat, though one should be careful not to confuse the definition here with the much more delicate global one discussed above. (For a survey on this topic see [31].)

We are going to look at asymptotically flat 4-manifolds which can be foliated by maximal hypersurfaces (i.e. hypersurfaces whose second fundamental form has zero trace). More precisely we want ${}^4M$ to be a product manifold $\mathbb{R} \times \underline{H}^3$, with $\underline{H}^3$ diffeomorphic to $\mathbb{R}^3$ and, with the metric given by

$$ds^2 = -\alpha^2(t,x)\, dt^2 + g_{ij}(t,x)\, dx^i\, dx^j \qquad (6)$$

where (x^1,x^2,x^3) are coordinates in $\underline{H}^3$. The hypersurfaces t = const. are space-like, asymptotically flat and maximal i.e.

<u>Maximal Foliation</u>: $\quad tr_g k := g^{ij}k_{ij} = 0$

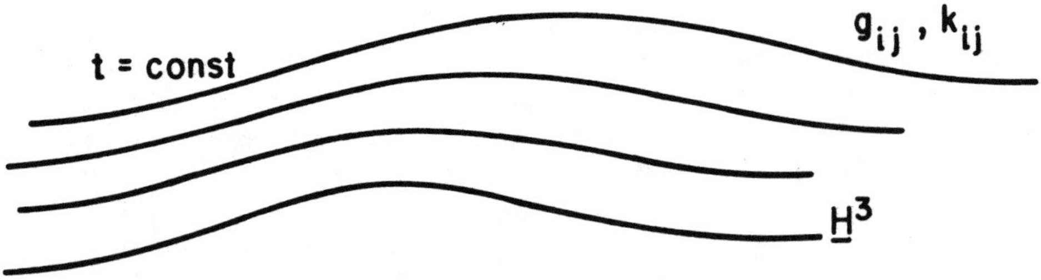

Figure 5

The possibility that the gauge condition (6), with tr k = 0, is the right one to use in the study of global space-times was suggested to us by the work of Bartnik [32].

Applying the Gauss and Gauss-Codazzi equations on each slice, we derive from (E.V.) the following

Constrained Equations. $\operatorname{tr} k = 0, \quad \operatorname{div} k = 0, \quad R = |k|^2$

Lapse Equation: $\Delta\alpha - |k|^2\alpha = 0$

Evolution Equations: $\frac{\partial g_{ij}}{\partial t} = -2\alpha k_{ij}$

$$\frac{\partial}{\partial t} k_{ij} = -\nabla_i\nabla_j\alpha + \alpha(R_{ij} - 2k_i^{\ell}k_{\ell j})$$

where Δ, div, tr and the operation of raising and lowering indices are taken relative to the induced metric g_{ij}, R is the scalar curvature of g_{ij}.

In particular we see from the constrained equations that $R \geq 0$. Now the Positive Mass Theorem of Schoen-Yau [6] and Witten [7] assures us that, in any coordinate system for which (5) holds, we have $M \geq 0$, and $M = 0$ if and only if the space is flat.

The above discussion allows us to give a more precise formulation of the problem we want to solve. We say that a triplet $(\underline{H}^3, g, k)$ formed by a 3-manifold $\underline{H}^3$ diffeomorphic to $\mathbb{R}^3$, a Riemannian metric g and a symmetric 2-tensor k is an "asymptotically flat" admissible Cauchy data if

$$\operatorname{tr}_g k = 0, \quad \operatorname{div}_g k = 0, \quad R = |k|_g^2$$

and there exists a coordinate system (x^1, x^2, x^3), in a neighborhood of infinity, such that

$$g_{ij} = \left(1 + \frac{2M}{r}\right)\delta_{ij} + o(r^{-3/2-\varepsilon}), \quad M > 0 \tag{7}$$

$$k_{ij} = o(r^{-5/2-\varepsilon}), \text{ as } r = \left(\sum (x^i)^2\right)^{1/2} \to \infty$$

and

$$\nabla^{\ell} g_{ij} = o(r^{-1-\ell}) \tag{7'}$$

$$\nabla^{\ell} k_{ij} = O(r^{-5/2-\varepsilon-\ell}) \text{ as } r \to \infty$$

for any partial derivatives of order $0 \leq \ell \leq 4$, and some small $\varepsilon > 0$.

Conjecture. Any system of asymptotically flat, small, admissible Cauchy data, leads to a globally smooth, complete solution (M^4, g) of the Einstein vacuum equations admitting a foliation (6) (with $\inf \alpha > \delta > 0$) and a smooth embedding of $\underline{H}^3$ to M^4 such that g_{ij}, k_{ij} are the first and second fundamental form of $\underline{H}^3$ to M^4.

In collaboration with D. Christodoulou we have developed a method which, if successful, will solve the conjecture by also deriving the correct asymptotic behavior of the Riemann curvature tensor along null and time-like directions. One essential part of our strategy is to use the Banchi's identities, which for an (E-V) space M^4 read

$$\nabla_{[\varepsilon} R_{\alpha\beta]\gamma\delta} = 0 \tag{8}$$

$$R_{\alpha\beta\gamma\delta}{}^{;\alpha} = 0 \ ,$$

as evolution equations and then derive g_{ij}, k_{ij}, α by solving purely elliptic problems. Indeed, let the dual of R be $^*R_{\alpha\beta\gamma\delta} = \frac{1}{2}\,\varepsilon_{\alpha\beta}{}^{\mu\nu} R_{\mu\nu\gamma\delta}$. Given $X \in \mathfrak{X}(M)$, define the 2-tensors

$$(ii_X R)_{\alpha\gamma} = R_{\alpha\beta\gamma\delta} X^\beta X^\delta, \qquad (ii_X^* R)_{\alpha\gamma} = {}^*R_{\alpha\beta\gamma\delta} X^\beta X^\delta \ .$$

Both $ii_X R$, $ii_X^* R$ or symmetric and trace-less and together determine the tensor R, provided that $\langle X,X\rangle \neq 0$. In analogy to the electromagnetic field, they can be called the <u>electric</u> and <u>magnetic</u> decomposition of R. In particular let $E = ii_T R$, $H = ii_T^* R$. Then

$$\begin{aligned} \mathrm{Ric} - k^2 &= E \\ \mathrm{curl}\ k &= H \end{aligned} \tag{9}$$

where Ric denotes the Ricci curvature of the slice, $(\mathrm{curl}\ k)_{ij} = \varepsilon_i{}^{st} k_{js;t}$. Therefore our strategy is to use the Bianchi identities (8) to estimate $R_{\alpha\beta\delta\gamma}$, with its electric and magnetic components E,H, and then use (9) together with the <u>constraint</u> and <u>lapse</u> equations to control k,g and α.

Notice that, on the initial slice,

$$E,H = O(r^{-3}) \quad \text{as} \quad r \to \infty \ .$$

However, note that the asymptotic behavior of the Lie

derivatives $\mathcal{L}_O E, \mathcal{L}_O H$ relative to the vector fields $O_1 = x_2 \partial_3 - x_3 \partial_2$, $O_2 = x_3 \partial_1 - x_1 \partial_3$, $O_3 = x_1 \partial_2 - x_2 \partial_1$ are better by a factor of r^{-1}. Thus we can verify that

$$\sum_{i=1}^{3} \sum_{\ell=0}^{s} \int (1+|x|)^{4+2\ell} \, |\nabla^{\ell} \mathcal{L}_{O_i} E(x)|^2 + |\nabla^{\ell} \mathcal{L}_{O_i} H(x)|^2) dx < \infty \tag{10}$$

on the initial slice, where s is some positive integer and ∇^{ℓ} the covariant derivative of order ℓ relative to the Riemann metric on the slice.

Given an Einstein space M^4 and a tensor $W_{\alpha\beta\delta\gamma}$ which satisfy all the symmetries of the Riemann tensor, i.e.

$$W_{\alpha\beta\gamma\delta} = - W_{\beta\alpha\gamma\delta} = W_{\gamma\delta\alpha\beta}$$

$$W_{\alpha\beta\gamma\delta} + W_{\alpha\gamma\delta\beta} + W_{\alpha\delta\beta\gamma} = 0$$

and also

$$W^{\alpha}{}_{\beta\alpha\delta} = 0$$

we say that it satisfies the <u>spin-2 equations</u> if

$$W_{\alpha\beta\gamma\delta;\varepsilon} + W_{\alpha\beta\delta\varepsilon;\gamma} + W_{\alpha\beta\varepsilon\gamma;\delta} = 0 \qquad (Sp_i)$$

$$W_{\alpha\beta\gamma\delta}{}^{;\delta} = 0 . \qquad (Sp_{ii})$$

Given a space-like hypersurface $\underline{H}^3$ with unit normal T we prescribe the initial conditions on $\underline{H}^3$

$$ii_T W = E_{(0)} , \quad ii_T^* W = H_{(0)}$$

with $ii_T W$, $ii_T^* W$ defined exactly as above for $R_{\alpha\beta\gamma\delta}$. The tensors $E_{(0)}$, $H_{(0)}$ are symmetric traceless in $\underline{H}^3$ satisfying the constraint equations

$$\text{div } E_{(0)} = 0 , \quad \text{div } H_{(0)} = 0 .$$

With the aim of understanding the asymptotic behavior of the curvature tensor R of solutions to the E-V equations it is natural to ask first the same question for the spin-2 equations (they are also called the linearized Einstein vacuum field equations) in Minkowski space, subject to initial conditions $E_{(0)}$, $H_{(0)}$ which verify the assumption (10). To achieve this we use a method which is similar to that mentioned above for solutions to (□), and which also works for (M), based on generalized Sobolev norms and energy estimates. The latter are connected to the concepts of Killing or conformal Killing vector fields and energy-momentum tensor which we discuss below. We say that

$$X \in \mathfrak{X}(M^{n+1}) \text{ is conformal Killing if}$$

$$X_{\mu;\nu} + X_{\nu;\mu} = \mathcal{L}_X g_{\mu\nu} = \chi\, g_{\mu\nu} \tag{11}$$

for some scalar χ. If $\chi = 0$ X is called simply, Killing.

Now define the following tensors corresponding to solutions u, F, W of (□), (M), (Sp),

$$Q_{\alpha\beta} = u_{,\alpha} u_{,\beta} - \frac{1}{2} g_{\alpha\beta}(g^{\mu\nu} u_{,\mu} u_{,\nu}) \qquad (Q_\Box)$$

$$Q_{\alpha\beta} = F_{\alpha\lambda}F_\beta{}^\lambda + {}^*F_{\alpha\lambda}{}^*F_\beta{}^\lambda \qquad (Q_M)$$

$$Q_{\alpha\beta\gamma\delta} = W_{\alpha\mu\beta\nu} W_\gamma{}^\mu{}_\delta{}^\nu + {}^*W_{\alpha\mu\beta\nu}{}^*W_\gamma{}^\mu{}_\delta{}^\nu . \qquad (Q_{Sp})$$

They are the energy momentum tensors of the field equations; $Q_{(Sp)}$ is the so-called Bell-Robinson tensor [33]. One can show that they are symmetric, traceless (with the exception of $(Q_\Box)$) in all pairs of indices and have zero divergence. Moreover they satisfy the <u>positive energy condition</u> which states that $Q(X_1,X_2)$ (or $Q(X_1,X_2,X_3,X_4)$ for $(Q_{(Sp)})$ is positive for all time-like vector fields X_1, X_2 (or X_1,X_2,X_3,X_4) with the same time-like orientation.

Now let X be an arbitrary vector field and Q the energy-norm tensor of either (□) or (M). Take

$$P^\mu = Q^{\mu\nu}X_\nu .$$

Then,

$$P^\mu{}_{;\mu} = \frac{1}{2} Q^{\mu\nu}(X_{\mu;\nu} + X_{\nu;\mu}) .$$

Consequently, if X is conformal Killing

$$P^\mu{}_{;\mu} = \frac{1}{2}\chi\, \mathrm{tr}(Q) . \tag{12}$$

In particular, if $\mathrm{tr}(Q) = 0$ (or X is Killing),

$$P^{\mu}{}_{;\mu} = 0 \;. \tag{12'}$$

The same holds true for $Q_{(Sp)}$ by considering $P^{\mu} = Q^{\mu\beta\gamma\delta} X_{\beta} Y_{\gamma} Z_{\delta}$ for any X,Y,Z conformal Killing. Assume that the divergence equation (12') holds true and that M^{3+1} is foliated by space-like hypersurfaces $t = \text{const}$ with future oriented unit normal T. Then, for (□) and (M)

$$\int_{H_t^3} Q(T,X) = \text{const} \quad \text{in } t \tag{13}$$

and for (Sp)

$$\int_{H_t^3} Q(T,X,Y,Z) \;=\; \text{const} \quad \text{in } t. \tag{13'}$$

By the positive energy condition the integrands are positive provided that X (and resp. X,Y,Z) are unit time-like, future oriented. Thus, to derive energy identities, with positive integrands, we need conformal Killing time-like vector-fields. Such vector fields exist in the Minkowski space to which we now restrict our attention.

First recall that the Minkowski space, introduced by H. Minkowski as the natural geometric space of Special Relativity, is the simplest example of an Einstein manifold i.e. the space diffeomorphic to $\mathbb{R}^{n+1}$ and with a canonical coordinate system $(x^0,x^1,\dots,x^n)$ for which

$$<\partial_{\mu}, \partial_{\nu}> = \eta_{\mu\nu} = \text{Diag}(-1,1,\dots,1) \;.$$

By abusing our previous notation we will denote x^0 by t and $x = (x^1,x^2,\dots,x^n)$. The Minkowski space possesses a large family of Killing and conformal Killing vector fields. The first one exhibited below are the generators of the groups of translations and Lorentz transformations (i.e. the Poincare group) and they are all Killing,

$$T_{\mu} = \partial_{\mu}; \qquad \mu = 0,1,\dots,n \tag{14_i}$$

$$L_{\mu\nu} = x_{\mu}\partial_{\nu} - x_{\nu}\partial_{\mu}; \qquad 0 \le \mu < \nu \le n \tag{14_{ii}}$$

with $x_{\mu} = g_{\mu\nu}x^{\nu}$. In particular for $i,j = 1,\dots,n$, $L_{ij} = x_i\partial_j - x_j\partial_i$ are the angular momentum operators corresponding to purely spatial rotations on the space-like slices t =

const. The vector fields $L_i = L_{i0} = x_i\partial_t + t\partial_i$ correspond to space-time rotations.

The next two groups of vector fields are conformal Killing and correspond to scaling transformations

$$S = x^\mu \partial_\mu \qquad (14_{iii})$$

and conformal transformations

$$K_\mu = -2x_\mu S + \langle S,S\rangle \partial_\mu . \qquad (14_{iv})$$

Denote by $\mathbb{K}$ the Lie algebra generated by ∂_μ, S, $L_{\mu\nu}$ for $0 \le \mu < \nu \le n$. Given any tensor W, we define

$$\begin{aligned} ||W(t)||^2_{\mathbb{K},k} &= \sum_{0\le\ell\le k} \int_{\mathbb{R}^n} |\mathcal{L}_{\Gamma_1}\cdots\mathcal{L}_{\Gamma_\ell} W(t,x)|^2\, dx \\ ||W||_{\mathbb{K},k} &= \sup_{t\ge 0} ||W(t)||_{\mathbb{K},k} \end{aligned} \qquad (15)$$

where $\Gamma_1,\ldots,\Gamma_\ell$ are any of the generators of $\mathbb{K}$ and $|\mathcal{L}_{\Gamma_1}\cdots\mathcal{L}_{\Gamma_\ell} W|$ is the Euclidian length, in $\mathbb{R}^{n+1}$, of the tensor $\mathcal{L}_{\Gamma_1}\cdots\mathcal{L}_{\Gamma_\ell} W$.

The linear vector fields S and $L_{\mu\nu}$ are dimensionless relative to scaling transformations $x \to \lambda x$ and span all directions in $\mathbb{R}^{n+1}$ except along the null cone

$-t^2 + (x^1)^2 + \ldots + (x^n)^2 = 0$

where S is null. In fact at every point in $\mathbb{R}^{3+1}$ we have the formulas

$$\partial_\mu = \langle S,S\rangle^{-1} (x_\mu S + x^\nu L_{\mu\nu}) \qquad (16)$$

for all $\mu = 0,1,\ldots,n$, where $\langle S,S\rangle = -t^2 + (x^1)^2 + \ldots + (x^n)^2$.

Similar formulas, which degenerate for $\langle S,S\rangle = 0$, express an electromagnetic tensor F in terms of $i_S F$, $i_S^* F$ and a spin-2 field W in terms of $ii_S W$, $ii_S^* W$. They imply in particular,

$$|F(t,x)| \leq c \frac{1}{|t-|x||} (|i_S F(t,x)| + |i_S^* F(t,x)|) \quad (16')$$

$$|W(t,x)| \leq c \frac{1}{|t-|x||^2} (|ii_S W(t,x)| + |ii_S^* W(t,x)|) \quad (16")$$

at all points $(t,x) \in \mathbb{R}^{3+1}$ with $t \neq |x|$.

The degeneracy of the formulas above along the null cone $<S,S> = 0$ is responsible for the different asymptotic behavior of solutions to field equations in Minkowski space, away from and near null directions. To describe what happens along the characteristic directions, it is useful to introduce the following null frame.

$$E_+ = E_{n+1} = \frac{1}{\sqrt{2}} (\partial_t + \Sigma_i \frac{x^i}{|x|} \partial_i) \quad (N)$$

$$E_- = E_n = \frac{1}{\sqrt{2}} (\partial_t - \Sigma_i \frac{x^i}{|x|} \partial_i).$$

Then pick $E_1, \ldots, E_{n-1}$ unit space-like vector fields orthogonal to each other and to E_+, E_-. Assuming them to be parallel transported in $\mathbb{R}^{n+1}$, we have

$$\nabla_{E_+} E_A = \nabla_{E_-} E_A = 0$$

for all $A = 1,2,\ldots,n-1$.

We are now ready to return to the field equations ($\Box$), (M), (Sp). Assume that the initial data $\psi_{(0)}, \psi_{(1)}$, for ($\Box$); $E_{(0)}, H_{(0)}$, for (M); and $E_{(0)}, H_{(0)}$, for (Sp), given at $t = 0$, satisfy the conditions

$$I_k^{(\Box)} (\psi_{(0)}, \psi_{(1)}) < \infty$$

$$I_k^{(M)} (E_{(0)}, H_{(0)}) < \infty$$

$$I_k^{(Sp)} (E_{(0)}, H_{(0)}) < \infty$$

where, $I^{(\Box)}$, $I^{(M)}$, $I^{(Sp)}$ are the weighted norms

$$I_k^{(\Box)} = [\sum_{\ell=0}^{k+1} \int_{\mathbb{R}^n} (1 + |x|^2)^\ell \, |\nabla^\ell \psi_{(0)}(x)|^2 \, dx \tag{17}$$

$$+ \sum_{\ell=0}^{k} \int_{\mathbb{R}^n} (1 + |x|^2)^\ell \, |\nabla^\ell \psi_{(1)}(x)|^2 \, dx]^{1/2}$$

$$I_k^{(M)} = [\sum_{\ell=0}^{k} \int_{\mathbb{R}^n} (1 + |x|^2)^{\ell+1} (|\nabla^\ell E_{(0)}(x)|^2 + |\nabla^\ell H_{(0)}(x)|^2)^2 dx]^{1/2} \tag{17'}$$

$$I_k^{(Sp)} = [\sum_{\ell=0}^{k} \int_{\mathbb{R}^3} (1+|x|^2)^{\ell+2} (|\nabla^\ell E_{(0)}(x)|^2 + |\nabla^\ell H_{(0)}(x)|^2) dx]^{1/2} \tag{17"}$$

with ∇^ℓ the ℓ-th spatial covariant derivatives on $t = 0$.

Now notice that among the vector fields displayed in (14) the only ones which are globally non space-like are T_0 and K_0. Since K_0 becomes null along $t^2 - |x|^2 = 0$ we replace it by the strictly time-like conformal Killing vector field

$$K = \partial_t + K_0 = (1 + t^2 + |x|^2)\partial_t + 2t\, x^i \partial_i \, . \tag{14_v}$$

Now let $X = T = T_0$ in the integrand in (13). Note that T_0 is the forward unit normal to the canonical foliation $t =$ const. We find,

$$Q_{(\Box)}(T,T) = \frac{1}{2}[(\partial_t u)^2 + \sum_i (\partial_i u)^2]$$

$$Q_{(M)}(T,T) = |i_T F|^2 + |i_T^* F|^2$$

which lead by integration to the usual energy identities for ($\Box$) and (M). In the same way, if we take $X = Y = Z = T$ in the integrand of (13') we infer that

$$Q_{(Sp)}(T,T,T,T) = |ii_T F|^2 + |ii_T^* F|^2 .$$

Now take $X = K$ in (13), we claim that

$$Q_{(M)}(T,K) = |i_S F|^2 + |i_S^* F|^2 + |i_T F|^2 + |i_T^* F|^2$$

with S the scaling vector field in (10_{iii}). This leads to

$$||i_S F(t)|| + ||i_S^* F(t)|| + ||i_T F(t)|| + ||i_T^* F(t)||$$

$$\leq I_0^{(M)}(E_{(0)}, H_{(0)}) \qquad (18_M)$$

for all $t \geq 0$. Similarly, taking X = T, Y = Z = K in (13') we find, for all $t \geq 0$

$$||ii_S W(t)|| + ||ii_S^* W(t)|| + ||ii_T W(t)|| + ||ii_T^* W(t)||$$

$$\leq I_0^{(Sp)}(E_{(0)}, H_{(0)}) \qquad (18_{Sp})$$

for all $t \geq 0$. By modifying somewhat the argument leading to (13'), to take into account the fact that $Q_{(\Box)}$ is not traceless, we derive, for solutions u of $(\Box)$ (see [34], [35])

$$||u(t)||_{\mathbb{K},1} \leq I_0^{(\Box)}(\psi_{(0)}, \psi_{(1)}) \qquad (18_\Box)$$

for any $t \geq 0$.

Notice that, together with (16), (16'), (16") the estimates (18), provide us already with some averaged decay properties of solutions to $(\Box)$, (M) and (Sp). Indeed one can show, in the particular case of $\Box u = 0$, that the local energy norm $\frac{1}{2} \int_\Omega [(\partial_t u)^2 + \sum_{i=1}^{n} (\partial_i u)^2]\, dx^1 dx^2 dx^3$ decays like t^{-2} for any compact set $\Omega \in \mathbb{R}^n$. These types of estimates were derived first by C. Morawetz [36], [37] for solutions of the wave equation in the exterior of a star shaped domain in $\mathbb{R}^n$. Similar local results were obtained by Costa-Strauss [38] for the Maxwell equations.

Next we note that if u,F,W are solutions the corresponding equations in Minkowski space and X is a generator of the Lie algebra $\mathbb{K}$ then the Lie derivatives $\mathcal{L}_X u$, $\mathcal{L}_X F$, $\mathcal{L}_X W$ are again solutions. This implies,

$$\|u\|_{\mathbb{K},k+1} \le C I_k^{(\square)}(\psi_{(0)},\psi_{(1)}) \tag{$19_\square$}$$

$$\|i_S F\|_{\mathbb{K},k} + \|i_S^* F\|_{\mathbb{K},k} + \|i_T F\|_{\mathbb{K},k} + \|i_T^* F\|_{\mathbb{K},k}$$

$$\le I_k^{(M)}(E_{(0)},H_{(0)}) \tag{19_M}$$

$$\|ii_S W\|_{\mathbb{K},k} + \|ii_S^* W\|_{\mathbb{K},k} + \|ii_T W\|_{\mathbb{K},k} + \|i_T^* W\|_{\mathbb{K},k}$$

$$\le I_k^{(Sp)}(E_{(0)},H_{(0)}) \ . \tag{19_{Sp}}$$

These generalized energy estimates are the key to our derivation of the asymptotic properties of solutions to the linear equations ($\square$), (M) and (Sp). To go from the weighted Sobolev norms displayed above to uniform norms we rely on the following (see [16], [34] and also [18], [39]).

<u>Global Sobolev Lemma</u>. Let $u = u(t,x)$ be a smooth scalar function in $\mathbb{R}^{n+1}$ with $x = (x^1,\ldots,x^n) \in \mathbb{R}^n$, $|x| = [\Sigma_i (x^i)^2]^{1/2}$. Assume $\|u\|_{\mathbb{K},k} < \infty$ for some $k > n/2$, with k a given positive integer. Then, given any generators $\Gamma_1,\ldots,\Gamma_\ell$ of $\mathbb{K}$, $0 \le \ell < k-n/2$, we have

$$|\Gamma_1 \cdots \Gamma_\ell \, u(t,x)|$$

$$\le C_k(1 + |t-|x||)^{-1/2}(1 + t + |x|)^{-(n-1)/2}\|u\|_{\mathbb{K},k}$$

uniformly in $t \ge 0$, $x \in \mathbb{R}^n$.

As a consequence of the G.S.L. and the formula (16) we derive

<u>Corollary</u>. Under the same assumptions as above

$$|\nabla^\ell u(t,x)|$$

$$\le C_k(1 + |t-|x||)^{-1/2-\ell}(1 + t + |x|)^{-(n-1)/2} \ \|u\|_{\mathbb{K},k}$$

uniformly for $t \ge 0$, $x \in \mathbb{R}^n$. Here $|\nabla^\ell u|$ denotes the euclidean norm in $\mathbb{R}^{n+1}$ of the ℓ-th covariant derivatives of u.

Finally, we state below the results which we can derive by using the estimates (18) and the G-S-L.

Theorem 1 ([34], [35]). Let u be a solution of (□) subject to the initial conditions

$$u = \psi_{(0)}\ , \quad u_t = \psi_{(1)}\ , \quad \text{at } t = 0$$

and assume that $I_k^{(\square)}(\psi_{(0)},\psi_{(1)}) < \infty$ for some $k > n/2$. Then, for all $0 \le \ell < k - n/2$,

$$|\nabla^{\ell} u(t,x)|$$

$$\le C_k(1 + |t - |x||)^{-1/2-\ell}(1 + t + |x|)^{-(n-1)/2}\ ,$$

$$\text{for } t \ge 0,\ x \in \mathbb{R}^n\ .$$

(ii) Moreover let $(E_1,\ldots,E_{n-1},E_n,E_{n+1})$ be the null frame discussed above, then

$$|E_{\mu_1} \ldots E_{\mu_\ell} u(t,x)| \le$$
$$\le C(1 + |t - |x||)^{-1/2-\ell_-}(1+t+|x|)^{-(n-1)/2-\ell+\ell_-}$$

for $t > 0$, $x \neq 0$ in $\mathbb{R}^n$, $|t-|x|| < 1/2(t+|x|)$ with ℓ_- the number of times $E_- = E_n$ appears in the sequence $E_{\mu_1},\ldots,E_{\mu_\ell}$.

Theorem 2 ([21]). Let F be a solution of the Maxwell equations (M), in $\mathbb{R}^{3+1}$, subject to the initial conditions

$$i_{\partial_t} F = E_{(0)}\ , \quad i^*_{\partial_t} F = H_{(0)} \text{ at } t = 0$$

where $E_{(0)}$, $H_{(0)}$ are vectors in $\mathbb{R}^3$ verifying the constraints $\operatorname{div} E_{(0)} = \operatorname{div} H_{(0)} = 0$ and such that $I_k^{(M)}(E_{(0)},H_{(0)}) < \infty$ for some integer $k > 3/2$. Then for every $0 \le \ell < k-3/2$

(i) $$|\nabla^{\ell} F(t,x)| \le C(1 + |t - |x||)^{-3/2-\ell}\ (1 + t + |x|)^{-1}$$

uniformly in $t \ge 0$, $x \in \mathbb{R}^3$.

Moreover, relative to the null frame $E_1,E_2,E_3= E_-$, $E_4= E_+$

(ii)
$$|F_{A3}(t,x)| \le C(1 + |t - |x||)^{-3/2}\ (1 + t + |x|)^{-1}$$
$$|F_{12}(t,x)| \le C(1 + |t - |x||)^{-1/2}\ (1 + t + |x|)^{-2}$$
$$|F_{34}(t,x)| \le C(1 + |t - |x||)^{-1/2}\ (1 + t + |x|)^{-2}$$
$$|F_{A4}(t,x)| \le C(1 + t + |x|)^{-5/2}$$

for all $A = 1,2$, $t \ge 0$, $x\neq 0$, in $\mathbb{R}^3$ and $|t-|x|| \le 1/2|t+|x||$.

Theorem 3 ([21]). Let W be a solution of the Spin-2 equations (Sp) subject to the initial conditions

$$i i_{\partial_t} W = E_{(0)}\,, \quad i i_{\partial_t} {}^{*}W = H_{(0)}\,,$$

for given symmetric, traceless 2-tensors $E_{(0)}$, $H_{(0)}$ in $\mathbb{R}^3$ verifying the constraints $\operatorname{div} E_{(0)} = \operatorname{div} H_{(0)} = 0$ and assume that $I_k^{(Sp)}(E_{(0)},H_{(0)}) < \infty$ for some integer $k > 3/2$. Then, for any $0 < \ell < k-3/2$,

$$\text{(i)} \quad |\nabla^{\ell} W(t,x)| \leq C(1 + |t - |x||)^{-5/2-\ell}\,(1 + t + |x|)^{-1}$$

uniformly in $t \geq 0$, $x \in \mathbb{R}^3$.

Moreover relative to the null frame E_1, E_2, E_3, E_4,

$$\begin{aligned}
\text{(ii)} \quad &|W_{A3B3}(t,x)| \leq C(1 + |t - |x||)^{-5/2}\,(1 + t + |x|)^{-1}\\
&|W_{A343}(t,x)| \leq C(1 + |t - |x||)^{-3/2}\,(1 + t + |x|)^{-2}\\
&|W_{1212}(t,x)| \leq C(1 + |t - |x||)^{-1/2}\,(1 + t + |x|)^{-3}\\
&|W_{1234}(t,x)| \leq C(1 + |t - |x||)^{-1/2}\,(1 + t + |x|)^{-3}\\
&|W_{A434}(t,x)| \leq C(1 + |t + |x||)^{-7/2}\\
&|W_{A4B4}(t,x)| \leq C(1 + |t + |x||)^{-7/2}
\end{aligned}$$

for all $A,B = 1,2$, $t \geq 0$, $x \neq 0$ in $\mathbb{R}^3$ and $|t-|x|| \leq 1/2(t+|x|)$.

In both Theorems 2, 3 the Lie derivatives relative to E_1, E_2, E_4 improve by a factor of $(1+t+|x|)^{-1}$ while the Lie derivative relative to E_3 improves only by $(1+|t-|x||)^{-1}$. However, the derivatives in the direction E_3 of certain components of F and W improve also by $(1+t+|x|)^{-1}$.

At the end of this lecture I want to discuss the relation between the initial condition $I_k^{(Sp)}(E_{(0)},H_{(0)}) < \infty$ and (10). In fact one can prove that, if $E_{(0)}$, $H_{(0)}$ are initial data for (Sp), satisfying $\operatorname{div} E_{(0)} = \operatorname{div} H_{(0)} = 0$ and (10), for some $s = k > 0$, then also $I_k^{(Sp)}(E_{(0)},H_{(0)}) < \infty$.* This is an

* This is not quite true for the Einstein Vacuum equations where one component of the curvature tensor will behave differently, due to the mass term in (7). Yet the arguments presented above can be modified to take the mass term into account.

interesting consequence of the Poincaré inequality on the unit sphere S^2 in $\mathbb{R}^3$. Moreover we note here that the initial asymptotic condition $I_k^{(Sp)}(E_{(0)}, H_{(0)}) < \infty$ is singular relative to the Penrose conformal compactification [30]. Thus the uniform decay rates for (Sp), which could be obtained in principle, by using that method, do not seem justified. The conformal compactification method was used by D. Christodoulou ([40], [41]) to prove global existence results for certain classes of nonlinear field equations for which the problem of slow decay of infinity of the initial data does not occur. A similar method was proposed by H. Friedrich [42] in an attempt to prove global existence results for the Einstein equations. He has been able to circumvent the difficulty of lack of conformal invariance of the Einstein equations and obtain a symmetric hyperbolic system for the transformed variables. The corresponding initial data are singular however at space-like infinity. The question whether that singularity propagates (in the radiative case) to a bona fide singularity of the compactified picture remains wide open, though certain recent results [43], [44] suggest that this is what happens.

The methods presented here have to be modified considerably for the (E-V) equations. In fact the mass term in the expansion of the induced metric on slices has the effect of distorting the asymptotic behavior of the light cones. To take this effect into account one has to modify the Killing and conformal Killing vector fields of the Minkowski space. This can be done [45] by a systematic study of "optical geometry" of a space-time whose curvature components behave similarly to those given by Theorem 3. Finally, to derive a global existence theorem for the Einstein equations one needs to treat the nonlinear terms which appear each time when we differentiate the Bianchi identities. For scalar wave equations it is known ([46], [47]) that quadratic nonlinearities could lead to formation of singularities for arbitrary small data unless the "null condition" is satisfied ([48], [40], [34], [35]). An appropriate version of the null condition, which takes into account the tensorial nature of the equations is satisfied by the Einstein vacuum equations.

REFERENCES

[1] S. Klainerman, G. Ponce, "global small amplitude solutions to nonlinear evolution equations," Comm. Pure Appl. Math. 36, 1983, 133-141.

[2] S. Klainerman, "Global Existence for Nonlinear Wave Equations," Comm. Pure Appl. Math. 33, 43-101, 1980.

[3] S. Klainerman, A. Majda, "Compressible and incompressible fluids," Comm. Pure Appl. Math. 35, 1982, 629-651.

[4] A. Majda, "Compressible Fluid Flow and Systems of Conservation Laws in Several Space Variables," Appl. Math. Sci. 53, Springer-Verlag.

[5] S. Klainerman, "Lecture Notes on Nonlinear Hyperbolic Systems," in preparation.

[6] R. Schoen-S. T. Yau, "On the proof of the positive mass conjecture in General Relativity", Comm. Math. Phys. 65, 1979, 45-76.

[7] E. Witten, "A simple proof of the positive energy theorem," Comm. Math. Phys. 80, 1981, 381-402.

[8] D. M. Eradley - V. Moncrief, "The global existence of Yang-Mills-Higgs fields in 4-dimensional Minkowski space, Comm. Math. Phys. 83, 171-191, and 193-212, 1982.

[9] K. Jörgens, Math. Z. 77, 295, 1961.

[10] W. von Wahl, "L^p decay rates for homogeneous wave equations," Math. Z. 120, 1971, pp. 93-106.

[11] I. E. Segal, "Dispersion of nonlinear relativistic equations," Ann. Sci. Ecole Norm. Sup. 4^e Ser., 459-497, 1968.

[12] W. Strauss, "Decay and asymptotics for $\Box u = f(u)$, J. Funct. Anal. 2, 1968, pp. 409-457.

[13] F. John, "Delayed singularity formation in solutions of nonlinear wave equations in higher dimensions," Comm. Pure Appl. Math. 29, 649-681, 1976.

[14] F. John, "Lower bounds for the life-span of solutions of nonlinear wave equations, in three dimensions," Comm. Pure Appl. Math. 36, 1-36, 1983.

[15] J. Shatah, "Global existence of small solutions to nonlinear evolution equations," J. Diff. Eqts. 78, 73-98, 1982.

[16] S. Klainerman, "Uniform decay estimates and the Lorentz invariance of the classical wave equation," Comm. Pure Appl. Math. 38, 321-332, 1985.

[17] F. John, S. Klainerman, "Almost global existence to nonlinear wave equations in three space dimensions," Comm. Pure Appl. Math. 37, 1984, 443-456.

[18] L. Hörmander, "On Sobolev spaces associated with some Lie algebras," Inst. Mittag-Leffler, Report No. 4, 1985.

[19] F. John, "A lower bound for the life-span of solutions of nonlinear wave equations in three space dimensions," preprint.

[20] L. Hörmander, "The life-span of classical solutions of nonlinear hyperbolic equations," Inst. Mittag-Leffler, report #5, 1985.

[21] D. Christodoulou-S. Klainerman, "Einstein geometry and linear field equations in Minkowski space," in preparation.

[22] Y. Choquet-Bruhat, "Theoreme d'existence pour certains systemes d'equations aux derivees partielles non lineaires," Acta Mathematica 88, 141-225,

[23] Y. Choquet-Bruhat, "Un theoreme d'instabilite pour certaines equations hyperbolique non lineaires," C. R. Acad. Sci. Paris 276A, 281.

[24] R. Geroch, "Asymptotic structure of space-time, P. Esposito and L. Witten (eds.) Plenum, New York, 1976.

[25] E. T. Newman - K. P. Todd, "Asymptotically Flat space-times," General Rel. and Gravitation, Vol. 2, A. Held, Plenum, 1980.

[26] H. Bondi, M. G. J. van der Burg, A. W. K. Metzner, "Proc. R. Sol, London Series, A 269, 21, 1962.

[27] H. Bondi, Nature 186, 535, 1960.

[28] R. K. Sachs, "Gravitational waves in general relativity VIII. Waves in asymptotically flat space-times," Proc. R. Soc. A270, 103-126 (1962).

[29] R. Penrose, "Proc. R. Soc. London Ser. A 284, 159, 1965.

[30] R. Penrose, "Conformal treatment of infinity," In Relativity, Groups and Topology, B. DeWitt and C. DeWitt (eds.), Gordon and Breach, 1963.

[31] A. Ashtekar, "Asymptotic Structure of the Gravitational Field at Spatial Infinity in General Relativity and Gravitation," Vol. 2, edited by A. Held in Plenum Publ. Corp., 1980.

[32] R. Bartnik, "Existence of maximal surfaces in asymptotically flat space-times", Comm. Math. Phys. 94, 155-175, 1984.

[33] L. Bel, "Introduction d'un tensur du quatrieme ordre," Computes Rendus 247, p. 1094, p. 1297, 1978.

[34] S. Klainerman, "The null condition and global existence to nonlinear wave equations," Lectures on Apl. Math., Vol. 23, 293- 326, 1986.

[35] L. Hörmander, "On global existence of solutions of nonlinear hyperbolic equations in $\mathbb{R}^{1+3}$," Inst. Mittag-Leffler Report No. 9, 1985.

[36] C. Morawetz, "The limiting amplitude principle," Comm. Pure Appl. Math. 15, 1962, pp. 349-362.

[37] P. Lax, R. Phillips, "Scattering theory for automorphic functions," Princeton Univ. Press, 1976.

[38] D. Costa, W. Strauss, "Energy Splitting," Quarterly of Appl. Math., 351-361, 1981.

[39] S. Klainerman, "Remarks on the global Sobolev inequalities in Minkowski space," to appear in Comm. Pure. Appl. Math.

[40] D. Christodoulou, "Global solutions of nonlinear hyperbolic equations for small data," to appear in Comm. Pure Appl. Math., 1986.

[41] D. Christodoulou, "Solutions globeles des equations des equations des champs de Yang-Mills," C. R. Acad. Sci. Paris, 293, Series A, pp. 39-42 (1981).

[42] H. Friedrichs, "On the hyperbolicity of Einstein and other field equations," Comm. Math. Phys. 100, 525, 1985.

[43] B. G. Schmidt, J. M. Stewart, "The scalar wave equation in a Schwarzschild space-time," Proc. R. Soc., London Series, A 367, 503, 1979.

[44] J. Porrill, J. M. Stewart, "Electromagnetic and gravitational fields in a Schwarzschild space-time," Proc. R. Soc., London Series, A 376, 451, 1981.

[45] D. Christodoulou, S. Klainerman, "Optical geometry and quasiconformal transformations of an Einstein space-time,"

[46] F. John, "Blow-up for quasilinear wave equations in three space dimensions," Comm. Pure Appl. Math. 34, 29-51, 1981.

[47] F. John, "Blow-up of radial solutions of $\Box u = \frac{\partial}{\partial t} F(u_t)$", preprint.

[48] S. Klainerman, "Long-time behavior of solutions to nonlinear wave equations," Proc. of Int. Congress for Mathematicans, Warsaw, 1982.

The work in this paper was supported by the National Science Foundation, Grant No. DMS-8504033.

S. Klainerman
Courant Institute of Mathematical Sciences,
New York University
251 Mercer Street
New York, New York 10012

Recent Progress on First Order Hamilton-Jacobi Equations

P. L. Lions

1. INTRODUCTION.

Hamilton-Jacobi equations have been studied for a long time : these classical (fully) nonlinear first-order equations were first introduced in the context of the Calculus of Variations. The most trivial example being the distance $d(x)$ between x and a fixed point (say in a Banach space, or a Riemannian manifold...) which is expected to solve $|\nabla d| = 1$ (defining appropriately $|\cdot|$ and where ∇ denotes the gradient). This also explains the relevance of Hamilton-Jacobi equations to various fields of Mathematical Physics and Mechanics. The interest in such equations has been renewed by Bellman's dynamic programming method which associates to optimal (deterministic) control problems involving, say, ordinary differential equations certain Hamilton-Jacobi equations (often referred to Bellman or Hamilton-Jacobi-Bellman equations in the engineering literature). A similar connection exists between deterministic differential games and the so-called Isaacs' equations. References are given below.

A very general formulation of Hamilton-Jacobi equations (HJ in short) is given by

$$H(x,u,\nabla u) = 0 \qquad \text{in } \Omega \qquad \text{(HJ)}$$

where (for example) Ω is an open set of a Banach space V, H (the Hamiltonian) is a given function uniformly continuous on bounded sets of $\Omega \times \mathbb{R} \times V^\star$ and $V^\star$ is the dual of V.

ISBN 0-12-195255-X

The unknown function u is scalar and ∇u denotes its (Fréchet) differential. Of course (HJ) will be supplemented with appropriate boundary conditions if $\Omega \neq V$ or conditions at infinity if $\Omega = V$.

It is well-known that such nonlinear first-order equations do not possess in general smooth (say C^1) solutions : one may again consider the example of the distance function or one may observe that if $u(x,t)$ $(x \in \mathbb{R}\ ,\ t \geq 0)$ "solves"

$$\frac{\partial u}{\partial t} + H\left(\frac{\partial u}{\partial x}\right) = 0 \qquad \text{on} \quad \mathbb{R} \times (0,\infty)$$

then $v(x,t) = \frac{\partial u}{\partial x}(x,t)$ "solves" the standard one-dimensional scalar conservation law

$$\frac{\partial v}{\partial t} + \frac{\partial}{\partial x}(H(v)) = 0 \quad \text{on} \quad \mathbb{R} \times (0,\infty)$$

where shocks (i.e. discontinuities of v) are well-known to form.

Thus, in the classical case $V = \mathbb{R}^N$, locally Lipschitz solutions solving a.e. (HJ) were considered by various authors (A. Douglis [11], S.N. Kruzkov [19], W.H. Fleming [13], [14]...) in an attempt of obtaining a global theory for (HJ) (or of understanding Bellman's or Isaacs'equations...). Global existence results were obtained in this class but with a tremendous loss of uniqueness : we refer to M.G. Crandall and P.L. Lions [6] for various such examples. Some particular uniqueness results in the case when $H(x,t,p)$ is convex in p were obtained by adding a very strong condition (semiconcavity in A. Douglis [11], S.N. Kruzkov [19] ; semi-superharmonicity in P.L. Lions [21]).

Recently, first in the case $V = \mathbb{R}^N$, M.G. Crandall and the author [6] introduced a new notion of solutions called viscosity solutions : this notion is recalled in section 2 below. Roughly speaking the idea is to approximate (HJ) via the vanishing viscosity method and assuming that the approximate solutions converge uniformly on compact subsets of Ω (as it is the case in all references above concerned with existence of locally Lipschitz solutions) the inequalities defining viscosity solutions are automatically inherited by the limit functions. And it was also shown in [6] that viscosity solutions are unique under quite general assumptions ; examples of existence results were given in [6] and more general ones in P.L. Lions [21] , [22] . It was observed in M.G. Crandall, L.C. Evans and P.L. Lions [4] that substantial improvements in the proofs may be obtained by appropriate choices of one of the many equivalent formulations of viscosity solutions.

It turned out that many questions on Hamilton-Jacobi equations and their applications may be solved by the use of viscosity solutions and we will not even attempt to review this incredibly fast growing field. Let us mention only that the use of viscosity solutions enables one to justify completely the dynamic programming approach in optimal control problems (cf. P.L. Lions [31]) or in differential games (cf. P.E. Souganidis [23], L.C. Evans and P.E. Souganidis [12]).

Our goal here is to explain what are viscosity solutions and why they are unique. After presenting the definitions we will concentrate on existence and uniqueness questions, saying a few words on the methods of proofs. And if the ideas and results below are taken from M.G. Crandall and P.L. Lions [7], [8] , [9] , M.G. Crandall, H. Ishii and P.L. Lions [5] , we would like to mention that several authors contributed to a better understanding of these questions including G. Barles [2] , [3] , P.E. Souganidis [24] , H. Ishii [15] , [16] , [17] , R. Jensen [18] ... Let us finally mention the reference [10] where a complete self-contained treatment of HJ via viscosity solutions is given.

As it will be seen below we will work in the context of possibly infinite dimensional spaces V : motivated by optimal control problems and the work of V. Barbu and G. Da Prato [1], M.G. Crandall and the author [9] introduced various methods to work with viscosity solutions in (quite) general Banach spaces. To simplify matters, we will work with "nice" Banach spaces V i.e. we will always assume V has the Radon-Nykodim property (RNP in short) and that V has a differentiable norm - like function $N(x)$ satisfying for some $C \geq 1$

$$\begin{gathered} N(x) \geq |x| \quad \text{on } V , \quad N \text{ is differentiable on } V-\{0\} \\ N(x) \leq C|x| \quad \text{on } V \end{gathered} \tag{0}$$

The role of the geometry of V may be seen on the example of $d(x) = |x|$ (norm of x) we mentioned in the beginning. And even if the RNP may be disposed of by the notion of strict viscosity solutions (see [9]), we do not know how to get rid of (0) thus excluding from our theory interesting spaces such as L^1, L^∞, C ...

2. DEFINITIONS OF VISCOSITY SOLUTIONS.

Let φ be any continuous function in Ω, we recall the definitions of the super differential of φ at $x \in \Omega$ (resp. the sub differential) denoted by $D^+\varphi(x)$ (resp. $D^-\varphi(x)$).

$$D^+\varphi(x) = \left\{ p \in V^\star \,/\, \limsup_{y \to x,\, y \in \Omega} \{\varphi(y)-\varphi(x)-(p,y-x)\} \cdot |y-x|^{-1} \leqslant 0 \right\}$$

(resp.

$$D^-\varphi(x) = \left\{ p \in V^\star \,/\, \liminf_{y \to x,\, y \in \Omega} \{\varphi(y)-\varphi(x)-(p,y-x)\} \cdot |y-x|^{-1} \geqslant 0 \right\}$$

where we denote by $|\cdot|$ the norms on V and $V^\star$ and $(\cdot,\cdot)$ is the duality pairing.

Definition. Let $u \in C(\Omega)$, u is a viscosity subsolution (resp. supersolution) of (HJ) if

$$\forall\, x \in \Omega ,\ \forall\, p \in D^+u(x) ,\quad H(x,u(x),p) \leqslant 0 \tag{1}$$

(resp.

$$\forall\, x \in \Omega ,\ \forall\, p \in D^-u(x) ,\quad H(x,u(x),p) \geqslant 0 \text{).} \tag{2}$$

And u is a viscosity solution of (HJ) if it is both a viscosity sub and supersolution of (HJ). ■

Using (0) for general spaces V, it is an easy exercise to show that the above is equivalent to (see [4], [6]).

Equivalent formulation : Let $u \in C(\Omega)$, u is a viscosity subsolution (resp. supersolution) of (HJ) if and only if we have for all $\varphi \in C(\Omega)$ differentiable on Ω the following inequality at any local maximum (resp. minimum) point x of $u-\varphi$

$$H(x,u(x),\nabla\varphi(x)) \leqslant 0 \tag{3}$$

(resp.

$$H(x,u(x),\nabla\varphi(x)) \geq 0 \quad) .$$

■

Remarks : In fact, we may add in the above statement that the local maximum point x is strict in the following sense

$$\sup \{ (u-\varphi)(y) \ / \ |y-x|=t \} < (u-\varphi)(x) \quad \text{for} \quad t > 0 \quad \text{small.}$$

And we may replace $\varphi \in C(\Omega)$ differentiable on Ω by $\varphi \in C^{1,1}_{loc}(\Omega)$ or $C^1(\Omega)$ if V is a Hilbert space (use for example the approximations in J.M. Lasry and P.L. Lions [20]), and even $\varphi \in C^\infty(\Omega)$ if $V = \mathbb{R}^N$. ■

The above formulation together with the preceding remarks yields the

1st Consistency property : Let $u_n \in C(\Omega)$ be a viscosity subsolution of

$$H_n(x,u_n,\nabla u_n) = 0 \qquad \text{in} \quad \Omega \tag{5}$$

where $H_n \in C(\Omega \times \mathbb{R} \times V^\star)$. We assume that H_n, u_n converge locally uniformly to some H,u i.e. for all $x \in \Omega$, $t \in \mathbb{R}$, $p \in V^\star$ there exists $\delta > 0$ such that

$$\sup \Big\{ |H_n(y,s,q)-H(y,s,q)|+|u_n(y)-u(y)| \ / \tag{6}$$
$$|y-x| \leq \delta \ , \ |s-t| \leq \delta \ , \ |p-q| \leq \delta \Big\} \underset{n}{\to} 0 .$$

Then u is a viscosity subsolution of (HJ). ■

We conclude this section by an elementary property of viscosity solutions in the classical case $\underline{V = \mathbb{R}^N}$ namely continuous functions obtained by the vanishing viscosity method are indeed viscosity solutions (thus justifying the terminology).

2nd Consistency property : Let $u_n \in C^2(\Omega)$ be a solution of

$$H_n(x,u_n,\nabla u_n) - \varepsilon_n \Delta u_n = 0 \qquad \text{in} \quad \Omega \tag{7}$$

for some sequence $\varepsilon_n > 0$, where $H_n(x,t,p) \in C(\Omega \times \mathbb{R} \times \mathbb{R}^N)$. We assume ε_n converges to 0 and H_n, u_n converge uniformly on compact subsets of $\Omega \times \mathbb{R} \times \mathbb{R}^N$, Ω respectively to some H,u . Then u is a viscosity solution of (HJ). ■

The idea of proof is : i) if $u-\varphi$ admits a strict local maximum at $x_o \in \Omega$ where φ, say, $\in C^2(\Omega)$ then for n large $u_n-\varphi$ admits a local maximum at a point $x_n \underset{n}{\to} x_o$, ii) at a maximum point x_n of $u_n-\varphi$ we have $\nabla\varphi(x_n)=\nabla u_n(x_n)$, $\Delta u_n(x_n) \leq \Delta\varphi(x_n)$ and thus

$$H_n(x_n,u_n(x_n),\nabla\varphi(x_n)) \leq \varepsilon_n \Delta\varphi(x_n)$$

and we let n go to $+\infty$.

Open problem : Take $H_n \equiv H$ in (7) and assume that ∇u_n is bounded in $L^\infty(\Omega)$ uniformly in n, does ∇u_n converges a.e. to ∇u ? The answer is known to be positive only if H is strictly convex or concave in p (this is obvious since ∇u_n and $H(x,u(x),\nabla u_n(x))$ converge to ∇u and $H(x,u,\nabla u)$ in $L^\infty(\Omega)$ weak ★).

3. HOW TO PROVE UNIQUENESS OF VISCOSITY SOLUTIONS.

We will consider two particular cases of (HJ) namely a problem in stationary form

$$u + H(x,u,\nabla u) = 0 \quad \text{in } V \tag{SP}$$

or some Cauchy problem

$$\frac{\partial u}{\partial t} + H(x,t,u,\nabla u) = 0 \quad \text{in } V \times]0,T[\ . \tag{CP}$$

In this section, we wish to explain the method to prove uniqueness of viscosity solution of (SP) or (CP). Even if the basic ideas of that method can be found in [6], [4], the right way to look at the original proofs has gradually emerged and our proposition follows the one in M.G. Crandall, H. Ishii and P.L. Lions [5] (in fact it unifies [5] and M.G. Crandall and P.L. Lions [8]). See also the general reference [10].

We will restrict ourselves here to the special case $V = \mathbb{R}^N$ and we leave the adaptation to the general case to the reader : this is easily done by using the abstract minimization property of Radon-Nykodim spaces due to C. Stegall [25] as in M.G. Crandall and P.L. Lions [9]. Another simplification of the presentation will be achieved in this section by considering only (SP).

We will first assume

$H(x,t,s,p)$ is nondecreasing with respect to s, (H1)

for all $x \in V$, $t \in [0,T]$, $p \in V^\star$

where $H(x,s,p)$ in (SP) is simply considered as some time (t) independent Hamiltonian $H(x,t,s,p)$.

Let us then consider two viscosity solutions $u,v \in C(\mathbb{R}^N)$ of (SP). Our goal is to prove that under appropriate conditions $u \equiv v$. Easy examples (see [9], [8]) show that restrictions on the growth of u,v,H at infinity are needed. We will hide these assumptions under conditions which may seem uncheckable to the reader a priori but are in fact present behind all uniqueness results. These conditions will illustrate the mechanism involved in uniqueness proofs.

We begin with a couple of remarks : we first notice that comparing viscosity solutions and classical C^1 solutions is straightforward indeed if u is a viscosity solution of (SP), $v \in C^1(\mathbb{R}^N)$ is a classical (or equivalently viscosity) solution of (SP) and $|u-v| \to 0$ as $|x| \to \infty$, then let $x_o \in \mathbb{R}^N$ be such that

$$\max_{\mathbb{R}^N} |u-v| = |u-v|(x_o) > 0 \qquad (8)$$

and assume for instance $(u-v)(x_o) > 0$. Then, by definition of viscosity solutions we find

$$u(x_o) + H(x_o,u(x_o),\nabla v(x_o)) \leq 0$$

while

$$v(x_o) + H(x_o,v(x_o),\nabla v(x_o)) = 0 .$$

Subtracting these inequalities and using (H1), we easily reach a contradiction.

The second remark is that the function $u(x)-v(y)=w(x,y)$ is a viscosity solution of

$$w + H(x,u(x),\nabla_x w) - H(y,v(y),-\nabla_y w) = 0 \qquad (9)$$
$$\text{in } \mathbb{R}^N \times \mathbb{R}^N .$$

This is a straightforward consequence of definitions.

Now the idea to prove $u \equiv v$ is to compare w viscosity solution of (9) with appropriate C^1 supersolutions of (9) which roughly speaking vanish for $x = y$. For instance, to prove that $u \leq v$ (and thus by symmetry $u \equiv v$) we argue in two steps.

Step 1 : $u-v$ is bounded from above.

In order to do so, we assume there exists a nonnegative constant C_o such that we can find for all $\varepsilon \in (0,1)$ a constant $M_\varepsilon \geq 0$, a C^1 function w_ε^1 on $\bar{B}_{M_\varepsilon} \times \bar{B}_{M_\varepsilon}$ satisfying

$$(\star) \quad \begin{cases} w_\varepsilon^1 + \inf\limits_{t \geq 0} \left\{H(x,t,\nabla_x w_\varepsilon^1) - H(y,t,-\nabla_y w_\varepsilon^1)\right\} \geq 0 \quad \text{on } B_{M_\varepsilon} \times B_{M_\varepsilon} \\ w_\varepsilon^1 \geq u(x)-v(y) \quad \text{on } \partial\left(B_{M_\varepsilon} \times B_{M_\varepsilon}\right), \quad M_\varepsilon \to +\infty \quad \text{as } \varepsilon \to 0, \\ \limsup\limits_{\varepsilon \to 0} w_\varepsilon^1(x,y) \leq C_o, \quad \text{for all } x \in \mathbb{R}^N, \\ \qquad y \in \mathbb{R}^N, \ |x-y| \leq 1 . \end{cases}$$

Then using the above observations (and (H1)), it is elementary to deduce

$$(u(x)-v(y)) \leq w_\varepsilon^1(x,y) \qquad \text{for } x,y \in B_{M_\varepsilon}$$

and letting ε go to 0 we find

$$\sup_{\mathbb{R}^N} (u-v) \leq C_o . \tag{10}$$

Step 2 : $u-v$ is nonpositive.

We next assume that for all $\varepsilon \in (0,1)$ we can find constants $M_\varepsilon > 0$, $\delta_\varepsilon \in (0,1)$ and a C^1 function w_ε^2 on $Q_\varepsilon = \{(x,y) \in \mathbb{R}^N \times \mathbb{R}^N \ / \ |x| \leq M_\varepsilon , \ |y| \leq M_\varepsilon , \ |x-y| \leq \delta_\varepsilon\}$ satisfying

$$(\star\star) \quad \begin{cases} w_\varepsilon^2 + \inf\limits_{t \geq 0} \left\{H(x,t,\nabla_x w_\varepsilon^2) - H(y,t,-\nabla_y w_\varepsilon^2)\right\} \geq 0 \quad \text{on } Q_\varepsilon \\ w_\varepsilon^2 \geq C_o \quad \text{on } \partial Q_\varepsilon, \quad M_\varepsilon \to +\infty \quad \text{as } \varepsilon \to 0, \\ w_\varepsilon^2(x,x) \to 0 \quad \text{as } \varepsilon \to 0 \quad \text{for all } x \in \mathbb{R}^N . \end{cases}$$

Again, using the above observations and Step 1, it is elementary to deduce

$$(u(x)-v(y)) \leqslant w_\varepsilon^2(x,y) \qquad \text{on} \quad Q_\varepsilon$$

and letting ε go to 0 we conclude. ∎

Of course, the terrible form of (★),(★★) might give the reader the feeling that those are conditions impossible to check. In fact, it turns out that (★),(★★) cover many situations and are not so difficult to implement. We give such an example in the next section. Let us, however, immediately mention that if H is x-independent, choosing $w_\varepsilon^1, w_\varepsilon^2$ as functions of $(x-y)$ makes (★),(★★) easy to check.

4. EXAMPLES OF EXISTENCE AND UNIQUENESS RESULTS.

We will consider Lipschitz, differentiable, nonnegative functions ν,μ on V such that

$$\mu(x) \geqslant |x| \quad \text{for} \quad |x| \quad \text{large}, \quad \nu(x) \to +\infty \quad \text{as} \quad |x|\to\infty. \tag{11}$$

We also introduce a function $d(x,y)$ on $V \times V$ which is Lipschitz, nonnegative, differentiable for $x \neq y$ and such that

$$d(x,x) = 0 \quad \text{on} \quad V, \quad d(x,y) \geqslant |x-y| \quad \text{on} \quad V \times V. \tag{12}$$

Typical examples are $\mu(x) = (1+N^2)^{1/2}$, $\nu(x) = \text{Log}\,(1+N^2)$, $d(x,y) = N(x-y)$ (recall that N satisfies (0) and that $N(x) = |x|$ if V is an Hilbert space).

We will use the following assumptions

$$\sup\Big\{H(x,t,s,p)-H(x,t,s,p+\lambda\nabla\mu(x)) \;/\; x \in V, \; t \in [0,T], \; s \in \mathbb{R}, \; |p| \leqslant R, \; 0 \leqslant \lambda \leqslant R\Big\} < \infty \quad \text{for all} \quad R < \infty \tag{H2}$$

$$\sup\Big\{H(x,t,s,p)-H(x,t,s,p+\lambda\nabla\nu(x)) \;/\; x \in V, \; t \in [0,T], \; s \in \mathbb{R}, \; |p| \leqslant R\Big\} \to 0 \quad \text{as} \quad \lambda \to 0_+ \quad \text{for all} \quad R < \infty \tag{H3}$$

$$\operatorname{Inf}\Big\{H(x,t,s,\lambda\nabla_x d(x,y))-H(y,t,s,-\lambda\nabla_y d(x,y)) \;/ \tag{H4}$$

$$t\in[0,T],\ s\in\mathbb{R},\ \lambda\geq 0,\quad x,y\in V \text{ with } |x-y|<d_o$$

$$\lambda d(x,y)+d(x,y)\leq\varepsilon\Big\}\ \to\ 0 \quad \text{as } \varepsilon\to 0 ,$$

for some $d_o > 0$.

We begin with an existence and uniqueness result for (CP).

Theorem 1 : 1) Uniqueness. We assume (H1)-(H4). Let u,v be viscosity solutions of (CP) such that u,v are uniformly continuous in $x\in V$, uniformly for $t\in[0,T]$ and $(u-v)^+(x,t)$ converges uniformly to $(u-v)^+(x,0)$ as $t\to 0_+$ on bounded sets of V. Then we have

$$\sup_V (u-v)^+(\cdot,t)\leq \sup_V (u-v)^+(\cdot,0) \quad \text{for all } t\in[0,T]. \tag{13}$$

2) Existence. We also assume (H1)-(H4). Let φ be uniformly continuous on V, then there exists a unique viscosity solution of (CP) uniformly continuous in $x\in V$ uniformly for $t\in[0,T]$ and uniformly continuous on bounded sets of $V\times[0,T]$ satisfying the initial condition

$$u(x,0) = \varphi(x) \quad \text{for all } x\in V. \tag{14}$$

■

In the case of (SP) we will need the following additional assumption : there exists $G\in C^1(0,\infty)$, uniformly continuous on $[0,\infty)$, nonnegative and nondecreasing such that

$$H(x,s,G'(d(x,y))\nabla_x d(x,y))-H(y,s,-G'(d(x,y))\nabla_y d(x,y))$$

$$+\,G(d(x,y))\geq 0 \quad \text{for } x\neq y\in V,\ s\in\mathbb{R}. \tag{H5}$$

Theorem 2 : 1) Uniqueness. We assume (H1)-(H4). Let u,v be viscosity solutions of (SP) which are uniformly continuous on V. Then $u\equiv v$ on V.

2) Existence. We assume (H1)-(H5). Then there exists a unique viscosity solution of (SP) which is uniformly continuous on V. ■

Remarks : i) Other examples of existence and uniqueness results are given in [9], [8].

ii) If H is uniformly continuous on $V \times B_R$ ($\forall R < \infty$) uniformly for $t \in [0,T]$, $s \in \mathbb{R}$ then (H2),(H3),(H5) holds with $\mu = \nu = (1+N^2)^{1/2}$ and $d(x,y) = N(x-y)$. ■

The uniqueness part of, say, Theorem 2 is proved using the strategy described in section 3 : we introduce

$$w_\varepsilon^1(x,y) = A(1+d^2(x,y))^{1/2} + B\,\beta\left(\mu(x) - \frac{1}{\varepsilon}\right) + C$$

where A,B,C are large enough (depending only on u,v), $\beta \in C^\infty(\mathbb{R})$, such that $\beta(t) \equiv 0$ for $t \leqslant 0$, $\beta'(t) \geqslant 0$ on $\mathbb{R}$, $\beta'(t) \equiv 1$ for $t \geqslant 1$, and we take M_ε large enough. Then we may check (★) using (H2) and (H4).

Next, to check (★★) we set

$$w_\varepsilon^2(x,y) = C_\varepsilon(\varepsilon^2+d^2(x,y))^{\gamma(\varepsilon)/2} + \omega(\varepsilon)\nu(x)$$

where C_ε is large enough (but $C_\varepsilon\, \varepsilon^{\gamma(\varepsilon)} \to 0$ as $\varepsilon \to 0_+$), $\gamma(\varepsilon) \in (0,1)$ is small enough and $\omega(\varepsilon)$ is small enough. The precise choices have to be made in order to check (★★) using (H3), (H4) (see [9], [5] for more details). This requires somewhat technical but completely elementary considerations.

In fact, the uniqueness part also yields a priori estimates : bounds or local bounds, moduli of uniform continuity. Indeed, the upper bounds on $u(x)-u(y)$ by $w_\varepsilon^2(x,y)$ above yield an estimate of the modulus of uniform continuity of u. If $V = \mathbb{R}^N$, Ascoli's theorem then easily yields the existence part using the vanishing viscosity method to obtain existence results for a restricted but "dense" class of Hamiltonians and then passing to the limit using the above a priori estimates. In the infinite dimensional case, the proof is much more complicated since neither the vanishing viscosity method nor Ascoli's theorem are available. We argue in [9] by obtaining partial existence results using convenient differential games and then passing to the limit by delicate stability results which are based upon the ideas used in the uniqueness proofs.

REFERENCES

1. V. Barbu and G. Da Prato, Hamilton-Jacobi equations in Hilbert spaces, Pitman, London, 1983.

2. G. Barles, Existence results for first-order Hamilton-Jacobi equations. Ann. IHP, Anal. Non Lin., 1 (1984), p. 325-340.

3. ________, Remarques sur des résultats d'existence pour les équations de Hamilton-Jacobi du premier ordre. Ann. IHP, Anal. Non Lin., 2 (1985), p. 21-33.

4. M.G. Crandall, L.C. Evans and P.L. Lions, Some properties of viscosity solutions of Hamilton-Jacobi equations. Trans. Amer. Math. Soc., 282 (1984), p. 487-502.

5. M.G. Crandall, H. Ishii and P.L. Lions, Uniqueness of viscosity solutions revisited. Preprint.

6. M.G. Crandall and P.L. Lions, Viscosity solutions of Hamilton-Jacobi equations. Trans. Amer. Math. Soc., 277 (1983), p. 1-42. Announced in C.R. Acad. Sci. Paris, 292 (1981), P. 183-186.

7. ________________________, On existence and uniqueness of solutions of Hamilton-Jacobi equations. Nonlinear Anal. T.M.A., 1985.

8. ________________________, Remarks on the existence and uniqueness of unbounded viscosity solutions of Hamilton-Jacobi equations. Announced in C.R. Acad. Sci. Paris, 298 (1984), p. 217-220.

9. ________________________, Hamilton-Jacobi equations in infinite dimensions. Part I, J. Func. Anal., 62 (1985), p. 379-396 ; Parts II and III to appear in J. Func. Anal.. Announced in C.R. Acad. Sci. Paris, 300 (1985), p. 67-70.

10. ______________________________, Book in preparation.

11. A. Douglis, The continuous dependence of generalized solutions of nonlinear partial differential equations upon initial data. Comm. Pure Appl. Math., 14 (1961) p. 267-284.

12. L.C. Evans and P.E. Souganidis, Differential games and representation formulas for solutions of Hamilton-Jacobi-Isaacs equations. Ind. Univ. Math. J., 33 (1984), P. 773-797.

13. W.H. Fleming, The Cauchy problem for a nonlinear first order partial differential equation. J. Diff. Eq., 5 (1969), p. 515-530.

14. ______________, The Cauchy problem for degenerate parabolic equations. J. Math. Mech., 13 (1964), p. 987-1008.

15. H. Ishii, Uniqueness of unbounded solutions of Hamilton-Jacobi equations. Ind. Univ. Math. J., 33 (1984), p. 721-748.

16. _________, Remarks on the existence of viscosity solutions of Hamilton-Jacobi equations. Bull. Facul. Sci. Eng., Chuo Univ., 26 (1983), p. 5-24.

17. _________, Existence and uniqueness of solutions of Hamilton-Jacobi equations. Preprint.

18. R. Jensen, personnal communication and work in preparation.

19. S.N. Kruzkov, Generalized solutions of the Hamilton-Jacobi equations of Eikonal type. I. Math. USSR Sbornik, 27 (1975), p. 406-446.

20. J.M. Lasry and P.L. Lions : A remark on regularization in Hilbert spaces.

21. P.L. Lions, Generalized solutions of Hamilton-Jacobi equations. Pitman, London, 1982.

22. __________, Existence results for first-order Hamilton-Jacobi equations. Ric. Mat. Napoli, 32 (1983), p. 1-23.

23. P.E. Souganidis, PhD dissertation, Univ. of Wisconsin-Madison, Madison, 1983.

24. _______________, Existence of viscosity solutions of Hamilton-Jacobi equations. J. Diff. Eq., 56 (1985), p. 345-390.

25. C. Stegall, Optimization of functions on certain subsets of Banach spaces. Math. Anal., 236 (1978), p. 171-176.

Ceremade
Université Paris-Dauphine
Place de Lattre de Tassigny
75775 Paris Cedex 16

The Focusing Singularity of the Nonlinear Schrödinger Equation

B. LeMesurier, G. Papanicolaou, C. Sulem, and P.-L. Sulem

1. Introduction

The nonlinear Schrödinger equation (NLS)

$$i\frac{\partial \phi}{\partial t} + \Delta\phi + |\phi|^{2\sigma}\phi = 0 \tag{1}$$

$$\phi(0,x) = \phi_0(x),\ x \text{ in } R^d \tag{2}$$

arises in many physical problems as an amplitude equation in nonlinear waves. The main property that we wish to study here is its focusing singularity. That is, solutions of the equation that have a singularity at a finite time $t = t^*$. It is called the focusing singularity because (1) arises in nonlinear optics in beam propagation [ASK, ES, CGT, Su, T1] where $\sigma = 1$ and $d = 2$. The nonlinear Schrödinger equation arises also in plasma physics [Go, GRH, K, VPT, Z1, Z2, ZSh, ZSy], in water waves [AK, N] and elsewhere.

We shall first review briefly the well known main analytical properties of (1), existence, invariants, special solutions and so on. Then we will construct some singular solutions. It will become clear in later sections that our analytical understanding of singular solutions for (1) is at present insufficient to explain what we observe in the numerical calculations. A description of our numerical methods is given in sections 10-13. Our numerical results are presented and discussed in section 14.

DIRECTIONS IN PARTIAL
DIFFERENTIAL EQUATIONS

ISBN 0-12-195255-X

2. Invariants

If NLS has a solution ϕ with $|\phi|^{2\sigma+2}$ and $|\nabla\phi|^2$ integrable, then it is easily seen that

$$M(\phi(t,.)) := \|\phi\|_2^2 = \text{constant} \tag{1}$$

$$H(\phi) := \|\nabla\phi(t,.)\|_2^2 - \frac{1}{\sigma+1}\|\phi(t,.)\|_{2\sigma+2}^{2\sigma+2} = \text{constant} \ . \tag{2}$$

These are the "mass" and "energy" invariants, respectively. We also have the variance identity

$$\frac{1}{8}\frac{d^2}{dt^2}\int |x|^2|\phi(t,x)|^2 dx = H - \frac{d\sigma-2}{2\sigma+2}\|\phi(t,.)\|_{2\sigma+2}^{2\sigma+2} \ . \tag{3}$$

By a solution of NLS we mean a function $\phi(t,x)$ that is continuous in t with values in $H^1(R^d)$. The solution $\phi(t,x)$ satisfies the equation (1.1.1) in the usual integral equation sense, obtained when the nonlinear term is treated as an inhomogeneous term.

Local in time existence for NLS can be obtained by standard iteration methods [GV]. When $\sigma d < 2$, the subcritical case, global existence is obtained using the invariant (2) and the Sobolev inequality as we review in section 1.4. That singular solutions exist when $\sigma d \geq 2$ follows from (3) [Gl] and we review this briefly in section 1.5.

3. Symmetries

The nonlinear Schrödinger equation is invariant under various transformations of dependent and independent variables. Thus a solution $\phi(t,x)$ of NLS remains a solution if translated in space or time $\phi(t,x) \to \phi(t-t_0,x-x_0)$. In addition, NLS is rotation invariant, invariant under phase change $\phi \to \phi e^{i\alpha}$, α real, and Gallilean transformations

$$\phi(t,x) \to \phi(t,x-\xi t)\exp\left(i\left(\xi x - |\xi|^2\frac{t}{4}\right)\right) \ . \tag{1}$$

A transformation of particular interest to us is the scaling transformation

$$\phi(t,x) \to \lambda^{1/\sigma}\phi(\lambda^2 t,\lambda x) \tag{2}$$

that also leaves NLS invariant.

The transformations we have just described exhaust the Lie transformation group that leaves NLS invariant for general σ and d. This can be shown [MP] by extending to NLS the analysis carried out for the nonlinear heat equation [BC, O]. When $\sigma d = 2$, the critical case, there is one more symmetry that has the form

$$\phi(t,x) \to \frac{1}{a^{d/2}}\phi(\tau,\theta)e^{\frac{ia\dot{a}|\theta|^2}{4}} \tag{3}$$

where $a = b(t^* - t)$, $t < t^*$ a constant and

$$\tau = \int_0^t \frac{1}{a^2(s)}ds,\ \theta = \frac{x}{a}\ . \tag{4}$$

The scaling function a can be considered a function of t or τ. In (3) $\dot{a}$ is the derivative of a with respect to t.

A simple consequence of the symmetry (3) in the critical case $\sigma d = 2$ is the existence of exact solutions that are are singular at some finite time t^* as will be given explicitly in section 1.6.

4. Global existence in the subcritical case

When $\sigma d < 2$, the subcritical case, global in time existence can be obtained using (1.2.2) and the Sobolev inequality

$$\|f\|_{2\sigma+2}^{2\sigma+2} \le C_{\sigma,d}^{2\sigma+2}\ \|\nabla f\|_2^{\sigma d}\ \|f\|_2^{2+\sigma(2-d)}\ ,\ 0 < \sigma < \frac{2}{d-2}\ ,\ d \ge 2 \tag{1}$$

to get a global in time estimate for $\|\nabla\phi(t)\|_2$ [GV].

The subcriticality condition $\sigma d < 2$ arises because we can use Minkowski's inequality on the right side of (1) and the term $\|\nabla f\|_2^2$. In fact we have

$$\begin{aligned}\|\nabla\phi(t,.)\|_2^2 &= H(\phi_0) + \frac{1}{\sigma+1}\|\phi(t,.)\|_{2\sigma+2}^{2\sigma+2}\\ &\le H(\phi_0) + \frac{1}{\sigma+1}C_{\sigma,d}^{2\sigma+2}\|\nabla\phi(t)\|_2^{\sigma d}\|\phi(t)\|_2^{2+\sigma(2-d)}\\ &\le H(\phi_0) + \frac{1}{2}\|\nabla\phi(t)\|_2^2 + C'\|\phi(t)\|_2^{p'}\end{aligned}$$

for some constant C' and exponent p'. Since the L^2 norm of ϕ is constant by (1.2.1) and $H(\phi_0)$ is a constant, the desired global in t bound for $\|\nabla\phi(t)\|_2$ follows.

5. The variance argument for blowup

When $\sigma d \geq 2$, the critical (equality) and supercritical cases, there are singular solutions of NLS. The existence of such singular solutions follows from the variance identity (1.2.3) [CGT, T1, Gl].

Suppose that $H(\phi_0) < 0$ and $\sigma d > 2$. Then there is a first time t_0 such that

$$\int |x|^2 |\phi(t,x)|^2 dx \to 0 \text{ as } t \to t_0 \tag{1}$$

assuming the solution exists up to t_0 for otherwise we already have a singular solution. Then the uncertainty inequality

$$\|f\|_2^2 \leq \frac{2}{d} \|\nabla f\|_2 \, \|xf\|_2$$

and (1.2.1) imply that $\|\nabla\phi(t)\|_2 \to \infty$ as $t \to t_0$ if (1) holds. This is the basic blowup argument.

Actually, the numerical computations show that $\|\nabla\phi(t)\|_2 \to \infty$ as $t \to t^*$ with t^* smaller than t_0. Blowup occurs even when the variance of $|\phi(t,x)|^2$ is infinite. This means that the above argument is limited and probably misleading because it relies on a quantity that does not really control blowup.

6. The R profile

There are simple exact solutions of NLS of the form

$$\phi(t,x) = e^{it} R(|x|) \tag{1}$$

where $R(r)$ satisfies

$$R'' + \frac{d-1}{r} R' - R + R^{2\sigma+1} = 0, \; r > 0 \tag{2}$$

with

$$R'(0) = 0, \, R(\infty) = 0 \ .$$

One can show that for $d \geq 1$ and $\sigma > 0$ there is a positive solution of this equation $R(r) > 0$, the ground state: this will be denoted R_0. Uniqueness of the ground state has been proven in several cases by Coffman [C] and McLeod and Serrin [MS]; in particular when $d = 1$ or 2 and when $d = 3$, $\sigma < 2$.

As we saw in section 1.3, there are exact solutions of NLS that are singular, that is blow up at a finite time, in the critical case $d\sigma = 2$. In the 2-dimensional cubic Schrödinger equation where most of our calculations have been done, we have the solution

$$\phi(t,x) = \frac{1}{t^*-t} R_0\left(\frac{|x|}{t^*-t}\right) \exp\left(\frac{-i|x|^2}{4(t^*-t)}\right), \; 0 \le t < t^* \; . \tag{3}$$

The blowup rate is $(t-t^*)^{-d/2}$: such a rate is not seen in our numerical calculations or any previous numerical work that we are aware of.

We will also refer to the more general R profile, $R(r;\mu_\infty,c_\infty)$, which is the solution of the initial value problem

$$R'' + \frac{d-1}{r} R' - c_\infty R + R^{2\sigma+1} = 0, \tag{4}$$

$$R(0) = \mu_\infty, \; R'(0) = 0 \; .$$

For any c_∞ value, these solutions only go to zero at infinity for a discrete set of initial values μ_∞.

7. Singular solutions by perturbation

Although the solution (1.6.3) is not observed numerically, the idea of constructing singular solutions based on the ground state R_0 is useful. In the critical case $\sigma = 1$, $d = 2$ (or more generally $\sigma d = 2$) there is a theorem of M. Weinstein [We2] that supports this idea. He showed that NLS has global in time solutions for any ϕ_0 in $H^1(R^d)$ when

$$\|\phi_0\|_2 < \|R_0\|_2 \tag{1}$$

where R is the ground state solution of (1.6.2) in the critical case. On the other hand, there exist singular solutions with any value of $\|\phi_0\|_2 \ge \|R_0\|_2$. This result tells us in particular that the solution (1.6.1) is at a stability boundary of the critical NLS.

Zakharov and Synakh [ZSy] have a heuristic argument that gives singular solutions in the case $\sigma = 1$, $d = 2$ of the form

$$\phi(t,x) \approx \frac{C}{(t^*-t)^{2/3}} R_0\left(\frac{|x|}{(t^*-t)^{2/3}}\right) e^{iS(t,x)} \tag{2}$$

where C is a constant and $S(t,x)$ is a real function, the phase, that is suitably determined. Numerical results of Zakharov and Synakh and more recently in

[SSP] tended to support the 2/3 power law for blowup. We now know that our results with t much closer to t^* do not support this type of blowup behaviour.

Wood [Wo] proposes another expansion method leading to singular solutions of the form

$$|\phi(t,0)| \approx \left(\frac{\ln(1/(t^*-t))}{t^*-t}\right)^{1/2} . \tag{3}$$

He also argues formally that the blowup rate of (2) can not occur, but this is not conclusive because he makes an ansatz on the form of the solutions that is too restricted to admit solutions with that power law blowup rate. Solutions with the "log-correction" singularity (3) were also proposed earlier by Vlasov, Piskunova and Talanov [VPT]. Their argument is more heuristic while Wood's is systematic but still formal. Numerical results are cited in both papers and they tend to prefer (3) over (2).

We should also point out that singular similarity solutions of the cubic Schrödinger equation in two dimensions can be found that have blowup

$$|\phi(t,x)| \approx \frac{1}{(t^*-t)^{1/2}} P\left(\frac{|x|}{(t^*-t)^{1/2}}\right) . \tag{4}$$

Such solutions were considered by Talanov [T2] but were rejected both by Wood [Wo] and Zakharov and Synakh [ZSy] because they have infinite L^2 and H^1 norm so the invariants (1.2.1) and (1.2.2) are infinite. Although our results are not incompatible with the blowup rate in (4), they suggest the form

$$|\phi(t,0)| \approx \frac{F(t^*-t)}{(t^*-t)^{1/2}} \tag{5}$$

where $F(s)$ goes to infinity as $s \to 0$ more slowly than $(\ln 1/s)^\gamma$ for any $\gamma > 0$.

How are the numerical results to be understood in the context of possibly well defined analytic solutions of the forms in (2) and (3)? We believe that stability considerations may be needed to explain why the solutions (2) or (3) are not observed numerically.

8. Dynamic rescaling

The considerations of the previous section suggest that scaling can be used effectively to solve numerically NLS deep into the blowup regime. The direct numerical integration of NLS, without substantial grid modifications to adjust to blowup, have not gone past amplitudes that are five or ten times larger than the initial amplitude [SSP]. With an adaptive grid method Wood [Wo] reports numerical results with amplitude amplification of order 200. Vlasov, Piskunova and Talanov [VPT] report numerical results with amplification factor 1.4×10^3, also using an adaptive grid method. In both papers the numerical results support the blowup form (7.3). With dynamic rescaling, a method we introduced and is described in this section, we obtain amplification factors of order 10^{10} for the cubic nonlinearity and 10^5 for the quintic with spatial grids that contain 256 or occasionally up to 512 points. The details of the numerical method and results are given in sections 10-13 and 14, respectively.

Dynamic rescaling is as follows. Let $L(t)$ be a positive function that will be specified later and let

$$\xi = \frac{r}{L(t)}, \quad \tau = \int_0^t \frac{1}{L^2(s)} ds \tag{1}$$

$$\phi(t,r) = \frac{1}{L(t)^{1/\sigma}} u(\tau,\xi) \ .$$

We restrict attention to radially symmetric solutions where $r = |x|$. One finds easily that $u(\tau,\xi)$ satisfies the equation

$$i\frac{\partial u}{\partial \tau} + \frac{\partial^2 u}{\partial \xi^2} + \frac{d-1}{\xi}\frac{\partial u}{\partial \xi} + |u|^{2\sigma}u - ia(\tau)\left(\frac{u}{\sigma} + \xi\frac{\partial u}{\partial \xi}\right) = 0 \tag{2}$$

where a can be considered to be a function of t or of τ and

$$a(\tau) = L\frac{dL}{dt} = \frac{1}{L}\frac{dL}{d\tau} = \frac{d(\ln L)}{d\tau} \ . \tag{3}$$

The initial condition is $u(\xi,0) = L(0)^{1/\sigma}\phi_0(L(0)\xi)$ The invariants (1.2.1) and (1.2.2) take the following form under this scaling

$$\|\phi(t,.)\|_2^2 = L(t)^{d-2/\sigma}\|u(\tau,.)\|_2^2 \tag{4}$$

$$H(\phi(t,.)) = L^{-p}(t)H(u(\tau,.)) \ , \ p = 2+2/\sigma-d \ . \tag{5}$$

How can we choose $L(t)$ so that the solution u of (2) is well behaved: that is, has no blowup? The most obvious choice is $L(t) = |\phi(t,0)|^{-1}$, which previous numerical results suggest would give a u profile whose amplitude tends to a constant at each ξ. Numerically, however, this is too sensitive to local errors. A better idea is to take some integral functional on the spatial variable that for ϕ goes to infinity at blowup, and choose L so that for u its value is constant in τ. In particular, since the H^1 norm of ϕ blows up, one can use the quantity

$$G(u(\tau,.)) := \|\nabla u(\tau,.)\|_2^2 \tag{6}$$

and the rescaling condition

$$G(u(\tau,.)) = G_0 \text{ , a prescribed constant.} \tag{7}$$

Differentiating G with respect to τ and using (2) gives, after some manipulations,

$$G_\tau = apG - 2\sigma \int_0^\infty |u|^{2\sigma-2}\mathrm{Im}\left(u\frac{\partial \bar{u}}{\partial \xi}\right)^2 \xi^{d-1} d\xi \,. \tag{8}$$

Thus setting

$$a(\tau) = \frac{2\sigma}{pG_0} \int_0^\infty |u|^{2\sigma-2}\mathrm{Im}\left(u\frac{\partial \bar{u}}{\partial \xi}\right)^2 \xi^{d-1} d\xi \tag{9}$$

gives

$$G_\tau = ap(G - G_0) \,. \tag{10}$$

Finally one must choose the initial value $L(0)$ to give $G(0) = G_0$. To do this note that, analogous to (5), one has

$$G(\phi(t,.)) = L^{-p}(t) G(u(\tau,.)) \,. \tag{11}$$

Thus the rescaling of (7) is equivalent to choosing

$$L(t) = \left(\frac{G_0}{\|\nabla\phi(t,.)\|_2^2}\right)^{1/p} \tag{12}$$

and in particular the case $\tau = 0$ gives the desired initial value. One could also derive the expression (9) for $a(\tau)$ directly from (12): differentiate (12) with respect to t and then use (3) and NLS.

Equation (10) helps to understand the inherent stability of the constant G rescaling with blowup solutions. For $p > 0$ always, because of the limit on σ needed for the basic local existence theorem, and $a < 0$ follows if $L_t < 0$ which will tend to hold for any reasonable rescaling when ϕ is blowing up. Thus even if $G \neq G_0$ due perhaps to numerical errors, the above choice of a will give

$G \to G_0$. This is born out in practice: errors in G do not accumulate in the way that errors in the computed values of $M(\phi(t,.))$ do.

A further numerical observation should be mentioned now: $|u(\tau,0)|$ converges to a constant so that the rescaling is asymptotically equivalent to the one originally suggested and rejected. That is, $L(t)^{1/\sigma}|\phi(t,0)| \to$ constant. On the basis of this, various forms conjectured for the behaviour of $|\phi(t,0)|$ are stated in terms of $L(t)$ in the discussion below.

With the rescaling (1), the NLS has been transformed into the pair (2), (9) that are to be solved simultaneously for u and a. The solution $u(\tau,\xi)$ behaves nicely for all τ, presumably because of (7). As $\tau \to \infty$, $t \to t^*$ and $L(t) \to 0$. The behaviour of $a(\tau)$ as $\tau \to \infty$ determines the behaviour of $L(t)$ as $t \to t^*$. If for example $a(\tau) \to k$, a non-zero constant, then $L(t) \approx \tilde{k}(t^*-t)^{1/2}$. If $a(\tau) \approx -c\tau^{-\gamma}$ with c some non-zero constant and $0 < \gamma < 1$ then

$$L(t) \approx \frac{\tilde{c}(t^*-t)^{1/2}}{\ln^{\alpha}(1/(t^*-t))}\ , \ t \to t^* \tag{10}$$

where $\alpha = \gamma/2(1-\gamma)$ and $\tilde{c}$ is another constant. If $a(\tau) \approx -c(\ln\tau)^{-2\alpha}$ then as $t \to t^*$

$$L(t) \approx \frac{\tilde{c}(t^*-t)^{1/2}}{(\ln\ln(1/(t^*-t)))^{\alpha}}\ . \tag{11}$$

Here $\alpha > 0$ is a constant.

9. The Q profile

There are exact solutions of (8.2) with $-a(\tau) = a_\infty$, a positive constant, of the form

$$u(\tau,\xi) = e^{ic_\infty\tau}Q(\xi) \tag{1}$$

where $Q(\xi)$ satisfies the ordinary differential equation

$$Q'' + \frac{d-1}{\xi}Q' - c_\infty Q + |Q|^{2\sigma}Q + ia_\infty\left(\frac{Q}{\sigma} + \xi Q'\right) = 0\ . \tag{2}$$

We define $Q(r;\mu_\infty,c_\infty,a_\infty)$ as the solution of the initial value problem

$$Q(0) = \mu_\infty,\ Q'(0) = 0\ .$$

These solutions correspond to similarity solutions of NLS that have the form

$$\phi(t,r) = \frac{1}{\left(a_\infty(t^* - t)\right)^{1/2\sigma}} Q\left(\frac{r}{\left(a_\infty(t^* - t)\right)^{1/2}}\right) e^{\frac{ic_\infty}{a_\infty}\ln\left(\frac{t^*}{t^* - t}\right)} . \tag{3}$$

In the critical case one can argue that solutions (2) exist but do not have finite mass M, because of slow decay at infinity. This does not fit well with the fixed finite mass $\|u(\tau,.)\|_2$ in the critical case, implied by (2.2) and (8.4). It means that complicated nonuniformities occur at infinity in the large time limit. Further, the numerical results below indicate that blowup in the critical case leads to a solution with $a(\tau) \to 0$ as $\tau \to \infty$. Thus it seems that the form (3) is not approached asymptotically in the critical case.

In the supercritical case there is no existence result for bounded solutions of the equation (2). Indeed, it can be shown that no solutions exist in the natural space $H_1 \cap L^{2\sigma+2}$. However, numerical solution of the I.V.P. (2) with $a_\infty > 0$ indicate that generically solutions decay at infinity. The decay is too slow to give finite mass but seems to give finite G and H, and indeed $H = 0$.

In the supercritical case, (2.2) and (8.4) give $\|u(\tau,.)\|_2 \to \infty$ as $\tau \to \infty$ so the seemingly infinite mass of Q is here good rather than bad for the hypothesis that u tends asymptotically to Q in some sense. Indeed the numerical results below indicate that in the supercritical case blowup does have asymptotically the form (3).

10. The Numerical Scheme for the Transformed Equations

In this section we shall describe the numerical method we use to solve NLS in the dynamically rescaled form of equations (9.2) and (9.9). We comment briefly on stability, conservation of the L^2 norm of ϕ, and its accuracy in practice.

It is sometimes convenient to think of a as an integral functional on functions of ξ rather than as a function of τ. Thus we will use interchangeably the notation $a(\tau)$, $a(u)$ and $a(u(\tau,.))$. The equations can be written in the form

$$\frac{\partial u}{\partial \tau} = i\Delta u + S(u) \tag{1a}$$

where S denotes the non-linear terms:

$$S = i|u|^2 u + a(u)\left(\frac{u}{\sigma} + \xi\frac{\partial u}{\partial \xi}\right) \tag{1b}$$

and

$$a(u) = \frac{1}{G_0}\int_0^\infty |u|^{2(\sigma-1)} \operatorname{Im}\left(u\frac{\partial \bar{u}}{\partial \xi}\right)^2 \xi^{d-1} d\xi \ . \tag{1c}$$

Here

$$G_0 = \int_0^\infty |\nabla u|^2 \xi^{d-1} d\xi$$

is a constant as shown in section 9. Introducing the notation

$$u^n = u(n\delta\tau,.)$$

$$u_j^n = u(n\delta\tau,\xi_j)$$

for the discretizations of $u(\tau,\xi)$, the time differencing used is the following combination of the Crank-Nicholson implicit method on the Laplacian term and Adams-Bashforth on the others.

$$\left(\frac{\partial u}{\partial \tau}\right)^{n+\frac{1}{2}} = i(\Delta u)^{n+\frac{1}{2}} + S(u)^{n+\frac{1}{2}} \tag{2a}$$

where

$$\left(\frac{\partial u}{\partial \tau}\right)^{n+\frac{1}{2}} := \frac{u^{n+1} - u^n}{\delta\tau} \tag{2b}$$

$$(\Delta u)^{n+\frac{1}{2}} := \Delta^{(N)}\left(\frac{u^n + u^{n+1}}{2}\right) \tag{2c}$$

$$S(u)^{n+\frac{1}{2}} := \frac{3}{2}S(u^n) - \frac{1}{2}S(u^{n-1}) \tag{2d}$$

$$S(u^n) := i|u^n|^2 u^n + a(u^n)\left(\frac{u^n}{\sigma} + \xi\partial_\xi^{(N)} u^n\right) \ . \tag{2e}$$

This implicit differencing, of the leading order terms only, eliminates a severe stability limit on time step size while avoiding time dependence or non-linearity in the implicit problem to be solved. The matrix involved can therefore be LU decomposed a single time at the start of the run, speeding execution.

The spatial grid is produced by a further change of variable to y, on the unit interval:

$$\xi = \xi_0 \left(\frac{e^{Ky} - 1}{K} \right) \tag{3a}$$

$$\xi_j = \xi(y_j) \tag{3b}$$

$$y_j = j/N, \; j = 0,\ldots,N-1 \tag{3c}$$

and the numerical differentiation and integration methods are described below. (The transformation is chosen to flexibly cover the large ranges of ξ values needed when $L(t)$ becomes small, due to the relation $r_{\max} = \xi_{\max} L(\tau)$ between the right-most points of the respective grids).

11. Evaluation of the discretized ξ differential operators, $\Delta^{(N)}$ and $\partial_\xi^{(N)}$

The discrete differential operators are evaluated in four steps:

(i) Extrapolate the data $u_0,\ldots,u_{N-1}$ to fictitious points $u_{-2},u_{-1},u_N,u_{N+1}$, using the boundary conditions, as follows: At "infinity" use

$$u_N = u_{N+1} = 0 \; .$$

At the origin, use the evenness conditions

$$\frac{\partial u}{\partial \xi}(0) = \frac{\partial^3 u}{\partial \xi^3}(0) = 0 \tag{1}$$

and the values u_0,u_1,u_2 to fit a fourth order Taylor polynomial in y at the origin, whose values are used for the fictitious point values.

(ii) Compute first and second derivatives in y at every grid point by the standard five point centred fourth order accurate scheme (quartic interpolation).

(iii) Convert these values to ξ derivatives using changes of coordinates based on exact expressions for the needed derivatives of the function $\xi(y)$.

(iv) Compute the Laplacian

$$\Delta u = \frac{\partial^2 u}{\partial \xi^2} + \frac{d-1}{\xi} \frac{\partial u}{\partial \xi}$$

at ξ_j, $j \geq 1$ in the obvious way, and for $j = 0$ use the limit expression

$$(\Delta u)(0) = d \frac{\partial^2 u}{\partial \xi^2}(0) \; .$$

We have also used a spectral method for spatial discretization [MPSS]. The results are the same in both cases.

12. Evaluation of ξ integrals

Compute

$$\int_0^\infty f(\xi)\xi^{d-1}d\xi = \int_0^1 f(\xi(y))\xi(y)^{d-1}\frac{d\xi}{dy}dy \tag{1}$$

by applying Simpson's method to the y integral, setting the integrand to zero at each extreme point. This is clearly correct at $y = 0$, and is justified at $y = 1$ by the assumption of sufficiently fast decay of all functions occurring as $\xi \to \infty$.

13. Evaluation of L and t

These quantities can be recovered from the equations

$$\frac{d(\ln L)}{d\tau} = a(\tau) \tag{1}$$

$$\frac{dt}{d\tau} = L^2 \tag{2}$$

(1) is integrated by the "Adams-Bashforth" method used above for the non-linear terms.

(2) is integrated using the midpoint rule.

Comments on stability, accuracy and conservation of norms

The observed stability limits on the time step size are in general accord with the result

$$\delta\tau \le O\left(\frac{K}{N\|a\|_\infty}\right) = O(1/N)$$

got by discarding the implicitly handled Laplacian term from the equation, and replacing a by its maximum norm.

An instability in the spatial discretization (not helped by reducing $\delta\tau$) in the form of time oscillations, sometimes occurs when K is too large and the solution is not (yet) settled into the asymptotic blowup behaviour. They appear to be due to the asymmetry of the discretized ξ Laplacian operator, which increases with K. Within limits, other important aspects of the solution seem to be reasonably accurate after the oscillations die down. It seems therefore that the lower wave number components are being solved with a fair degree of accuracy even during the oscillatory stage.

14. Numerical Results

First some notation used below shall be introduced.

c denotes the τ derivative of the phase of u at the origin.

$$\alpha(\tau) := \frac{a(\tau)}{c(\tau)} \quad \text{and} \quad \mu(\tau) := \frac{|u(\tau,0)|}{\sqrt{c(\tau)}}$$

are normalized values of $a(\tau)$ and the absolute value at the origin: the values that would occur for a choice of G_0 giving $c(\tau) = 1$. α is the quantity called k in [MPSS], and observed there to have a universal limiting value of about -0.917 in the three dimensional cubic case.

α_∞ and μ_∞ denote (conjectured) limits as $\tau \to \infty$ of the previous quantities.

Gc denotes the Gaussian initial profile ce^{-r^2}.

Lc denotes the Lorentzian initial profile $\frac{c}{1+r^2}$.

It was commented earlier that as measures of the accuracy of a numerical solution run, both of the conserved quantities $G(u)$ and $M(\phi)$ have their failings. The former will be conserved numerically regardless of errors, so long as a is negative and still computed accurately, because of equation (9.9). The latter is sensitive to errors at large values of ξ that may not have significant effects on the other integrals (a and G) or local quantities at smaller ξ values as discussed in section 2.4.

Thus we usually give two sets of final data, based on a conservative and a more liberal test of accuracy, as follows. All runs are required of course to be free from manifest inaccuracies. Once this requirement is satisfied as best possible by grid refinement and adjustment of the grid point parameters r_0 and K, failure of M conservation is usually the first sign of inaccuracy, as mentioned earlier. When runs have been continued to a point where this inaccuracy but no other has appeared, the tables summarizing the runs give data at the end of the run and also at an earlier time at which all conserved quantities are still accurate. In such cases, graphs against τ continue to the later time, and so should be viewed with the above in mind.

We discuss explicitly only results for the case of a cubic nonlinearity, $\sigma = 1$. Similar calculations for $\sigma = 2$ in both critical and supercritical cases reproduce all the same qualitative features to be observed below, though with less accuracy due to the greater numerical difficulties produced by the sharper non-linearity. We start with a supercritical case as the results here offer more clear-cut interpretations, and are in accord with theoretically based expectations. They also provide a point of reference for the anomalous behaviour in the critical case.

14.1. Three dimensional (supercritical) case

Table 1 summarizes the results for three choices of initial data.

Table 1

Initial data	G6	L6	G8
$\delta\tau$	0.005	0.005	0.005
N	256	256	256
K	24	24	24
ξ_{max}	2.12×10^{10}	1.77×10^{10}	3.77×10^{9}
at $\tau = 0$			
r_0	5	5	5
r_{max}	5.02×10^{9}	5.02×10^{9}	5.02×10^{9}
M	5.6399	28.27	10.027
G	4.	4.	4.
L	0.2364	0.2829	0.13300
H	−18.975	−49.48	−83.362
at $\tau =$	19.	16.	26.
r_{max}	2.48	83.89	3.52
M	5.6308	28.03	10.002
G	4.0001	4.00008	4.0004
L	1.17×10^{-10}	4.72×10^{-10}	9.32×10^{-11}
t	0.05575392	0.06117872	0.03707588
a	−1.35709295	−1.3572322	−1.35676095
c	1.47970425	1.4797823	1.47961474
$\lVert u \rVert_\infty$	2.29353629	2.2936829	2.29316366
α	−0.917137969	−0.9171837	−0.91696907
μ	1.88546359	1.885534455	1.885214311
at $\tau =$	20.	20.	30.
r_{max}	0.064	0.037	0.0015
M	3.88	1.98	0.047
G	4.0001	4.00008	4.0004
L	3.02×10^{-11}	2.07×10^{-11}	4.10×10^{-13}
t	0.05575392	0.06117872	0.03707588
a	−1.35709300	−1.3572341	−1.35676107
c	1.47970429	1.4797837	1.47961484
$\lVert u \rVert_\infty$	2.29353632	2.2936840	2.29316374
α	−0.917137977	−0.9171841	−0.91696907
μ	1.88546362	1.885534429	1.885214315

These are similar to but smaller than the initial data studied in [MPSS], which were $G6\sqrt{2}$ and $L6\sqrt{2}$, but the qualitative features of the growth rate observed there are confirmed. In particular one sees the universal limit behaviour of α, and in addition of μ, with the limiting values being

$$\alpha_\infty \approx -0.9172 \ , \quad \mu_\infty \approx 1.8855 \ .$$

The three cases give qualitatively similar results, so graphs are displayed only for G6 data. Figure 1 is a plot of $a(\tau)$ against τ.

It has been observed in various studies, for example [SSP], that when solution profiles at different times are similarity rescaled to compensate for the blowup, they appear to converge to a limit function. As noted in section 9, this is conjectured to be a Q profile in the supercritical case. Indeed it has been reported in [MPSS] that when rescaled to $c = 1$, the limiting profile always has the same α_∞ value. It is also implicit in those results that the μ_∞ value will be universal, and we will now test and confirm these conjectures further.

The rescaling in the algorithm is sufficient to allow direct comparison of u profiles without a further rescaling to, say, constant supremum norm. Unless noted otherwise, we will refer only to the G6 run: the other cases have the same behaviour in all significant respects.

Figures (2a-c) show profiles at τ= 5, 10, 15 and 20 on several scales. The convergence is excellent from $\tau = 10$ on which, not surprisingly, is the time at which $a(\tau)$, $c(\tau)$ and $|u(\tau,0)|$ have become nearly constant.

Figures (3a,b) show the profile at τ=20, superimposed on the Q profile with parameters taken from the solution: $\mu_\infty = |u(20,0)|$, $c_\infty = c(20)$ and $a_\infty = -a(20)$. The numerical solution of the I.V.P. for Q sets the limit $\xi \leq 30$ by losing accuracy (the slight oscillations visible on the second curve are from this, not a real phenomenon), so the fit of the solution amplitude profile to the Q amplitude is excellent as far out as we can go.

This fit is very sensitive to the parameters of the Q profile: a 4% change in them increased the errors in the fit by a factor of 50.

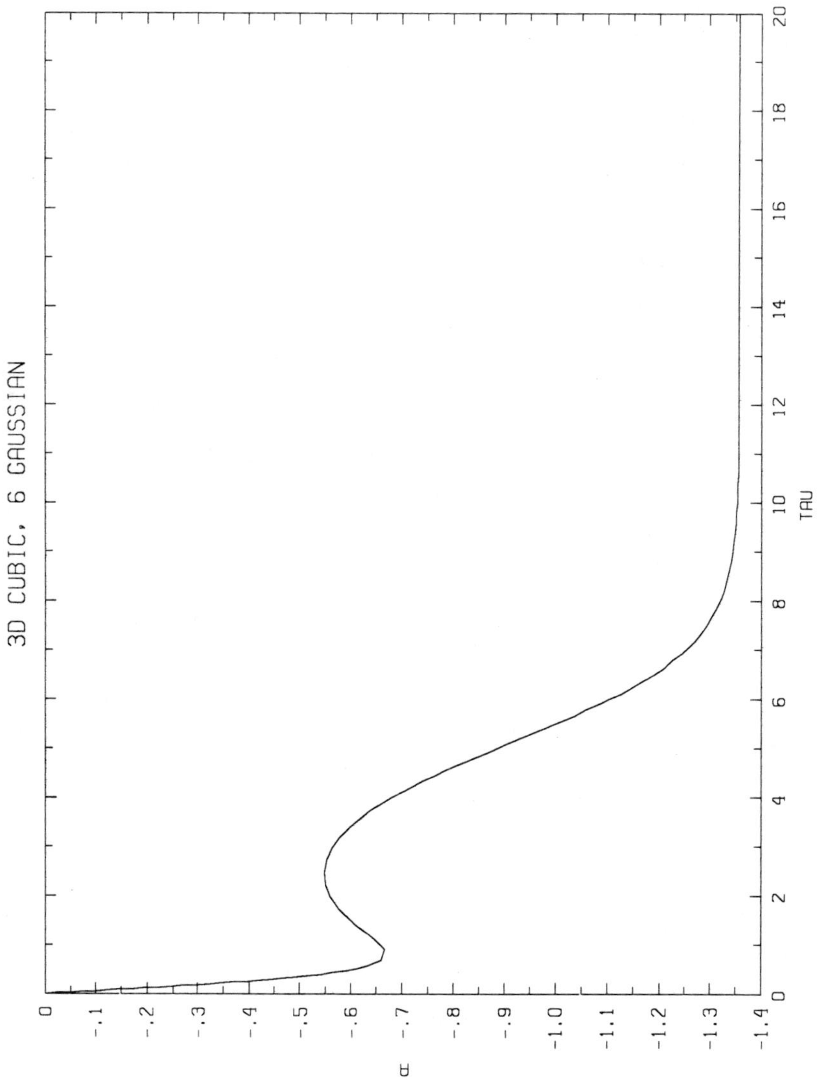
3D CUBIC, 6 GAUSSIAN
TAU
a
0
2
4
6
8
10
12
14
16
18
20
0
-.1
-.2
-.3
-.4
-.5
-.6
-.7
-.8
-.9
-1.0
-1.1
-1.2
-1.3
-1.4

Figure 1

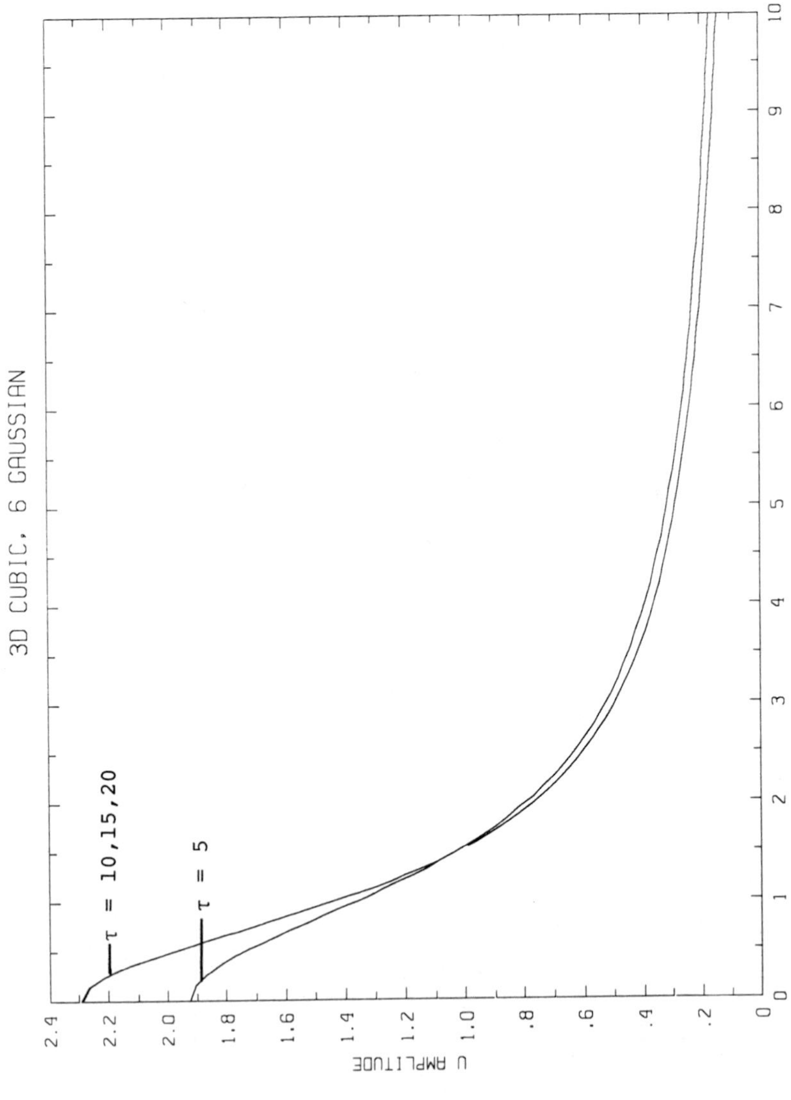
3D CUBIC, 6 GAUSSIAN
τ = 10,15,20
τ = 5
U AMPLITUDE
XI
2.4
2.2
2.0
1.8
1.6
1.4
1.2
1.0
.8
.6
.4
.2
0
0
1
2
3
4
5
6
7
8
9
10

Figure 2a

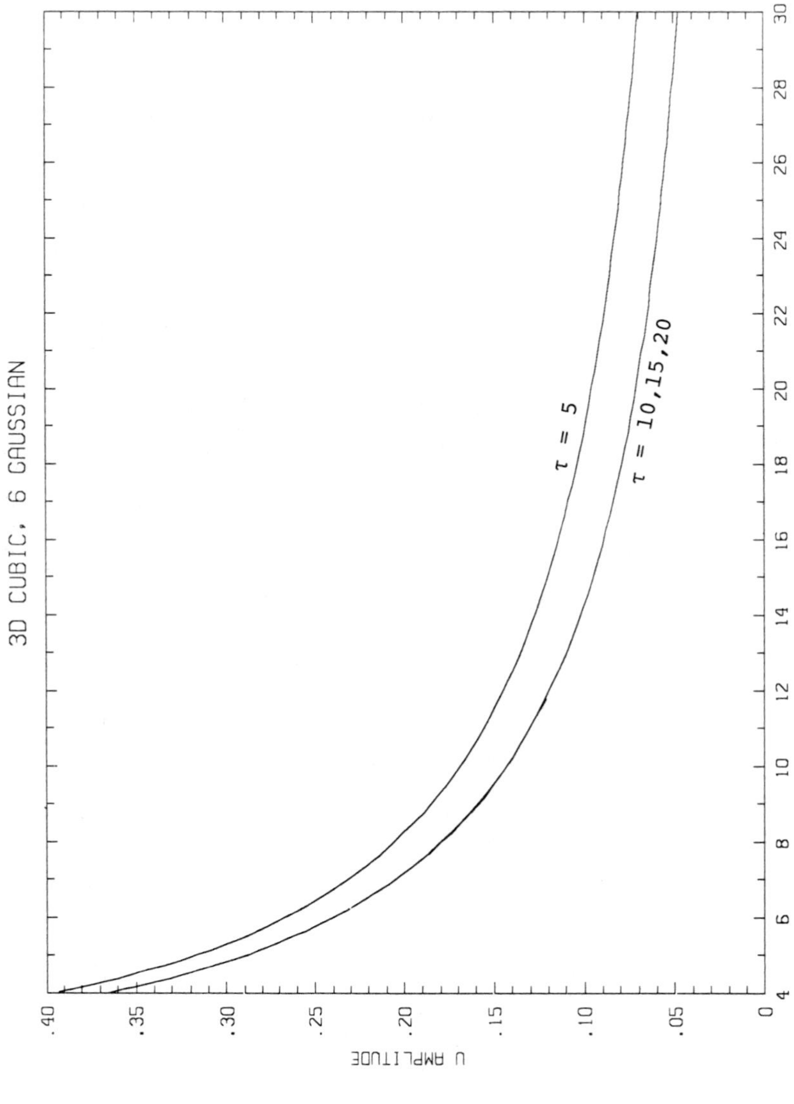
3D CUBIC, 6 GAUSSIAN
U AMPLITUDE
XI
τ = 5
τ = 10,15,20
0
.05
.10
.15
.20
.25
.30
.35
.40
4
6
8
10
12
14
16
18
20
22
24
26
28
30

Figure 2b

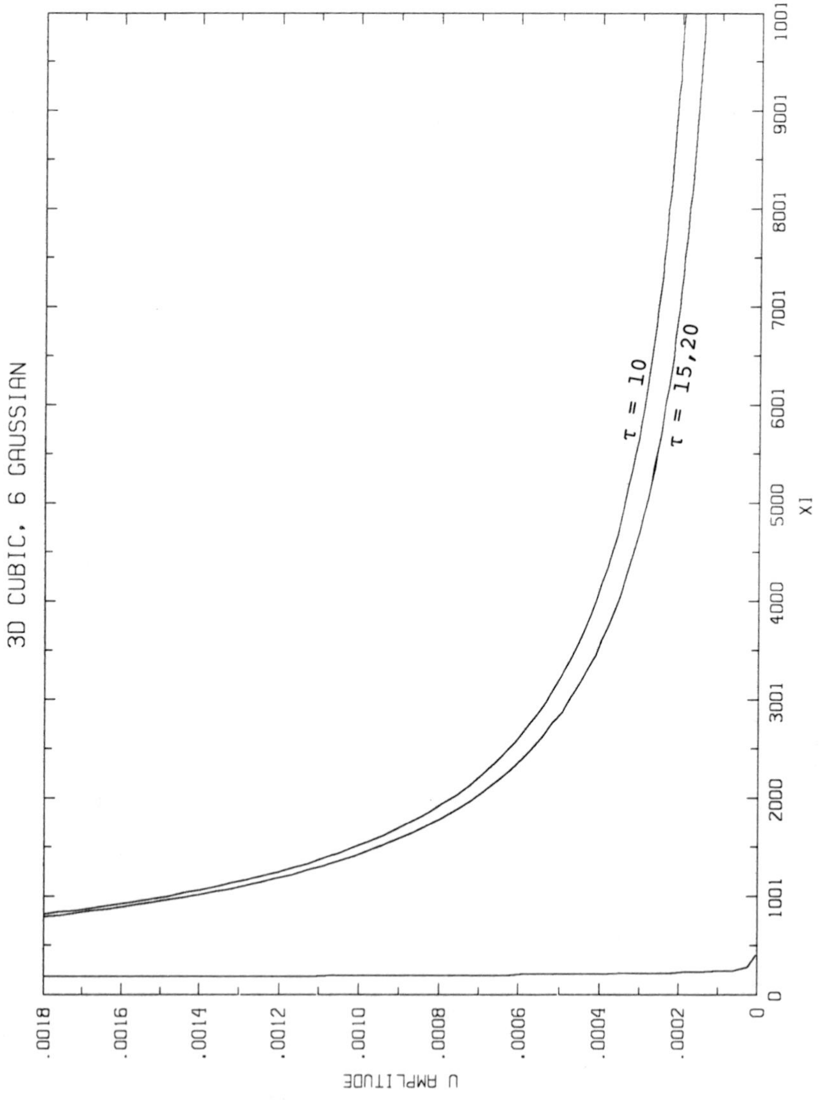
3D CUBIC, 6 GAUSSIAN
τ = 10
τ = 15,20
U AMPLITUDE
XI
0
.0002
.0004
.0006
.0008
.0010
.0012
.0014
.0016
.0018
0
1001
2000
3001
4000
5000
6001
7001
8001
9001
10010

Figure 2c

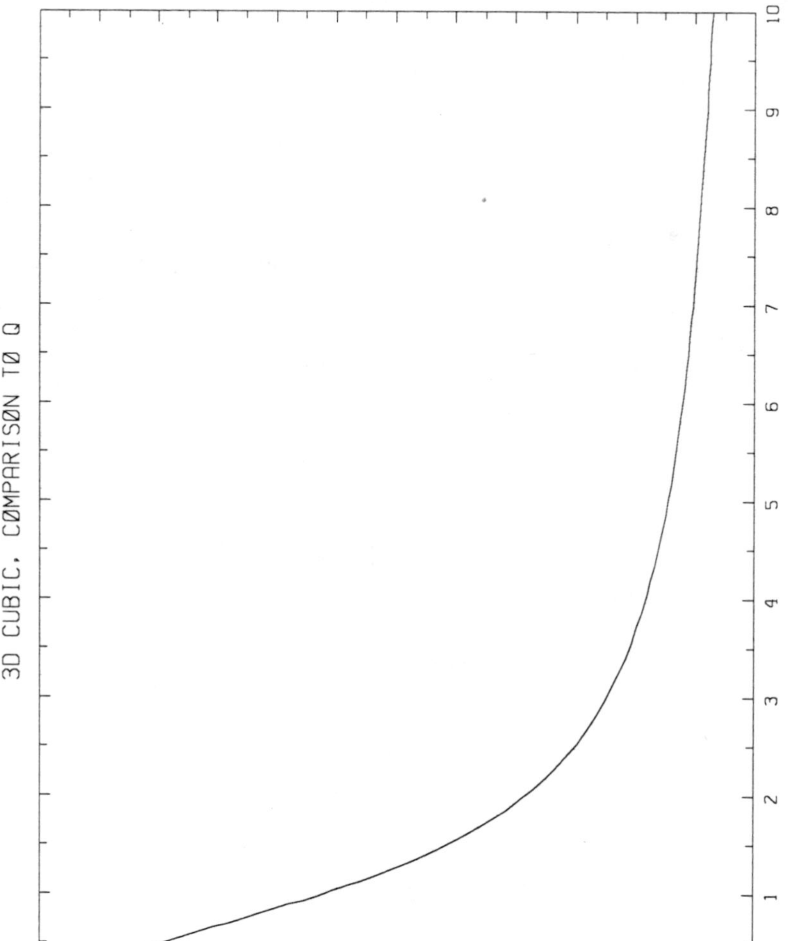
3D CUBIC, CØMPARISØN TØ Q
ABSØLUTE VALUE
XI
2.4
2.2
2.0
1.8
1.6
1.4
1.2
1.0
.8
.6
.4
.2
0
0
1
2
3
4
5
6
7
8
9
10

Figure 3a

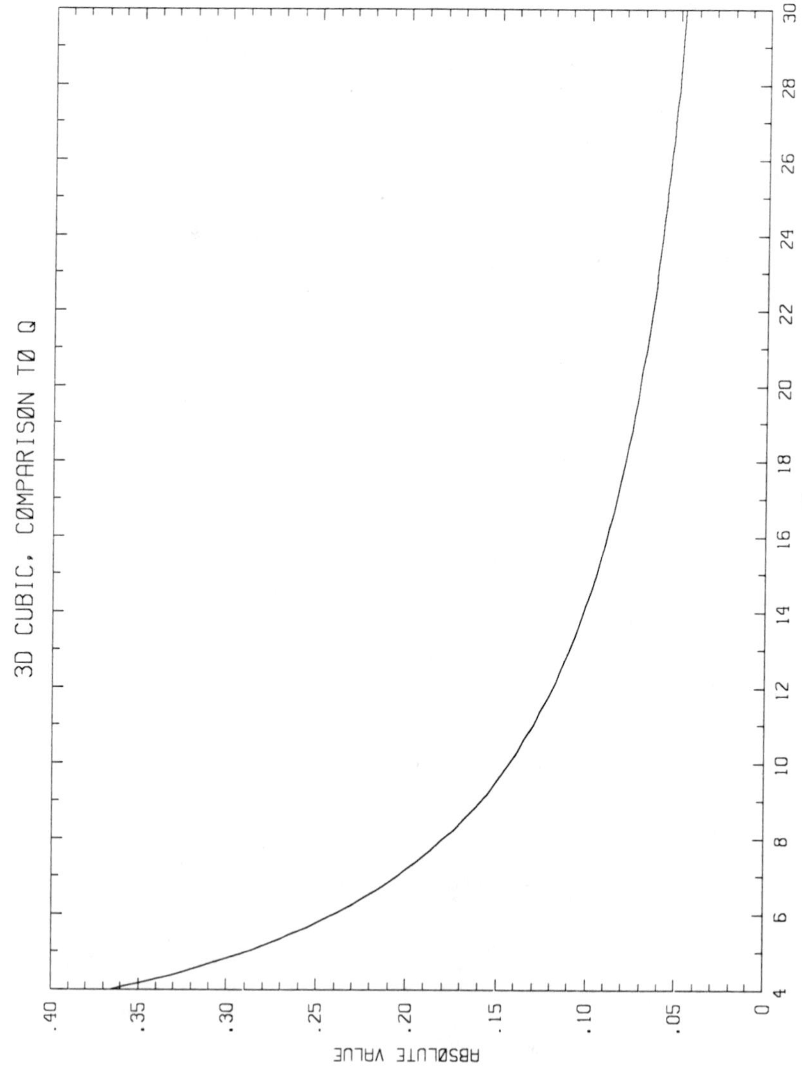
3D CUBIC, CØMPARISØN TØ Q
ABSØLUTE VALUE
0
.05
.10
.15
.20
.25
.30
.35
.40
XI
4
6
8
10
12
14
16
18
20
22
24
26
28
30

Figure 3b

14.2. Two dimensional (critical) case

Table 2 summarizes the results for four choices of initial data. As before the Lorentzian profile case is not covered by the blow-up theorem because of its infinite variance. The G2.77 and G2.8 cases have positive energy but M larger than the critical value of approximately 1.86 ([SSP]) so neither the global existence theorem of Weinstein [We1] or the blowup theorem apply. Nevertheless, both these runs indicate blowup of the solution.

Table 2

Initial data	G4	L4	G2.77	G2.8
$\delta\tau$	0.005	0.005	0.005	0.005
N	256	256	512	512
K	24	24	8	10
ξ_{max}	1.41×10^{10}	1.16×10^{10}	3591	21153
at $\tau = 0$				
r_0	5	5	5	5
r_{max}	5.02×10^{9}	5.02×10^{9}	1833	10684
M	4.	8.	1.918225	1.96
G	1.	1.	1.	1.
L	0.3536	0.433	0.5105	0.505076
H	$-8.$	$-16.$	$+0.157$	0.0078
at $\tau =$	–	–	60.	55.
r_{max}	–	–	8.1	4.7
M	–	–	1.909	1.95
G	–	–	1.000000	1.000000
L	–	–	2.26×10^{-3}	2.21×10^{-4}
t	–	–	1.677417	1.0474510
a	–	–	-0.09622	-0.143725
c	–	–	0.54217	0.548848
$\|\|u\|\|_\infty$	–	–	1.61206	1.605887
α	–	–	-0.1775	-0.26187
μ	–	–	2.1893	2.16765
at $\tau =$	100.	100.	100.	100.
r_{max}	6.15	8.816	0.17	0.008
M	3.985	7.997	1.88	1.91
G	1.000017	1.000009	1.000000	1.000000
L	4.33×10^{-10}	7.60×10^{-10}	4.824×10^{-5}	3.80×10^{-7}
t	0.14544861	0.16106060	1.67744358	1.0474512
a	-0.158849	-0.159053	-0.09624	-0.139550
c	0.551603	0.551632	0.54214	0.548148
$\|\|u\|\|_\infty$	1.603291	1.603250	1.61206	1.606550
α	-0.28798	-0.28833	-0.1775	-0.25458
μ	2.15874	2.15862	2.1894	2.16993

Final α and μ values are given to show that no non-zero universal value has manifested itself for the former, but does seem to be present for the latter. This is consistent with the conjecture that the limiting profile in the critical case will be the ground state function R_0 of section 1.6, so that it can be suggested that the limits will the parameters in the Q equation giving that solution:

$$\alpha_\infty = 0, \ \mu_\infty \approx 2.2052 \ .$$

The poorness of the approach to these proposed limits is one reflection of the far slower appearance of the large τ limiting behaviour in the critical case. Figures (4a-c) are the $a(\tau)$ against τ plots for the Gaussian profiles. The results for the L4 case are qualitatively the same as for G4 data.

The G4 case shows a pattern for $a(\tau)$ that does not fit with the previously conjectured scaling laws. Those would require, as noted above, a power law decay:

$$a(\tau)\tau^\beta \to C \ .$$

This seems approximately correct in the early stages, which cover the range of amplitudes seen in earlier work, but then the decay slows down substantially. In this respect it should be noted that in the G4 case the amplitude growth reaches a factor of about 1500 at $\tau = 26$ so effectively it is only the data up to this point that has been seen in the earlier work mentioned in section 8.

Figure (4b) for case G2.77 is even more unexpected: it shows all the characteristics of the supercritical case. That is, two turning points, then convergence to a non-zero constant. Figure (4c) for G2.8 is qualitatively the same, but shows a visible slow decrease in $|a|$, consistent with the idea that one just has very slow decay to zero. Unfortunately there are still very small oscillations in the G2.77 solution that make it difficult to seek a decreasing trend. Thus a second run was done with G2.77 initial data, on a more extended spatial grid to allow the run to continue until $\tau = 160$. This had $\delta\tau = .002$, $K = 14$, and r and N as before. It suffered from the oscillations mentioned above, particularly around $\tau = 20$, but they died down substantially by $\tau = 60$, and the data after this time agreed well with the previous run. When one looked beyond $\tau = 100$, the oscillations in $a(\tau)$ continue to decay (to the fifth significant figure and below), and a clear slow decrease in magnitude is visible. For $120 \leq \tau \leq 160$, a_τ is almost constant at 0.0007.

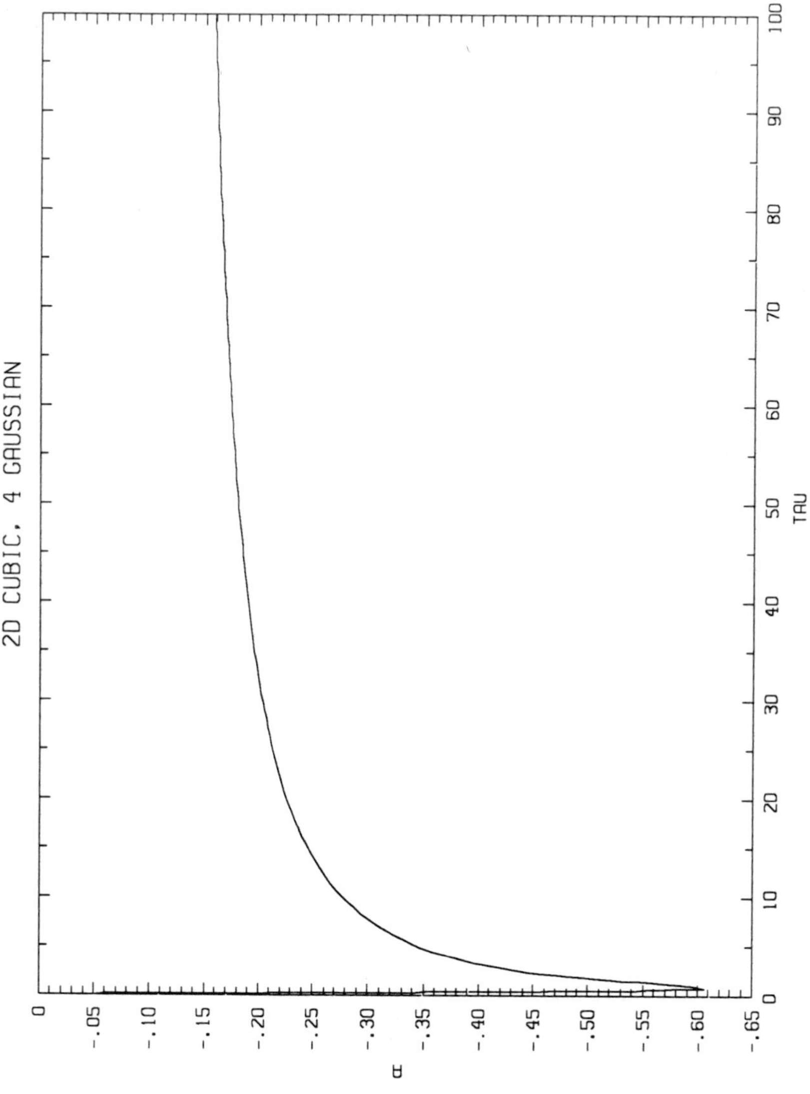
2D CUBIC, 4 GAUSSIAN
TAU
α
0
10
20
30
40
50
60
70
80
90
100
0
-.05
-.10
-.15
-.20
-.25
-.30
-.35
-.40
-.45
-.50
-.55
-.60
-.65

Figure 4a

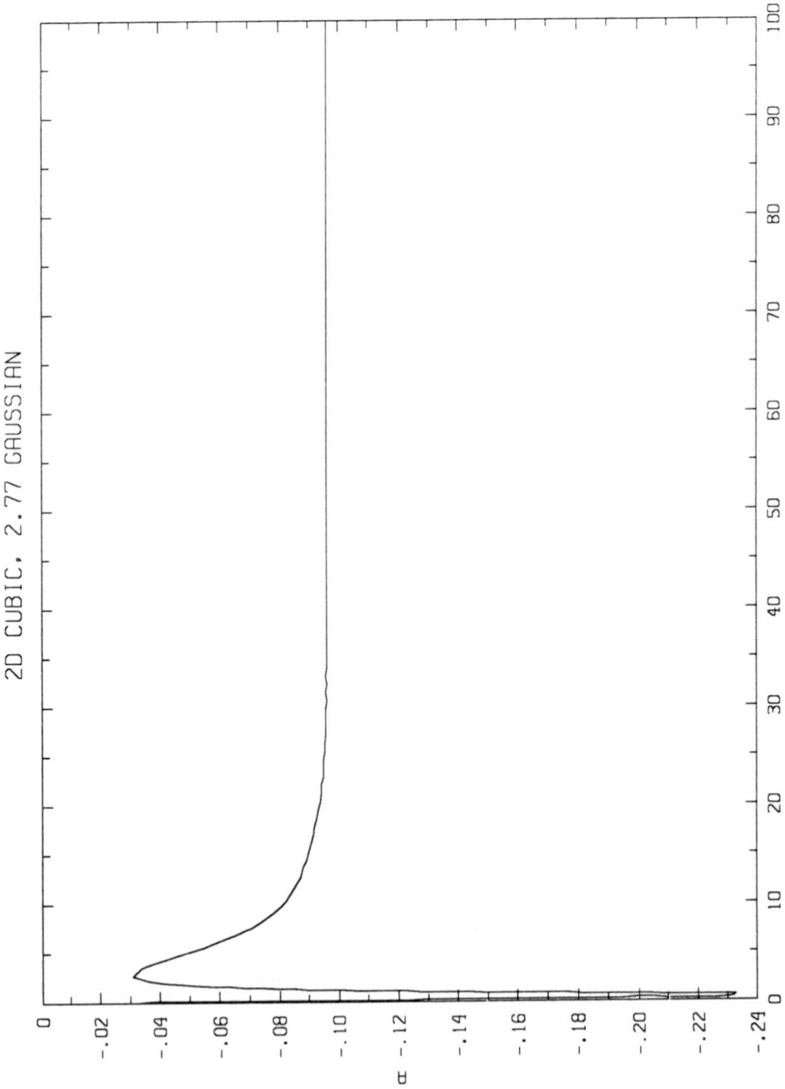
2D CUBIC, 2.77 GAUSSIAN
TAU
α
0
10
20
30
40
50
60
70
80
90
100
0
-.02
-.04
-.06
-.08
-.10
-.12
-.14
-.16
-.18
-.20
-.22
-.24

Figure 4b

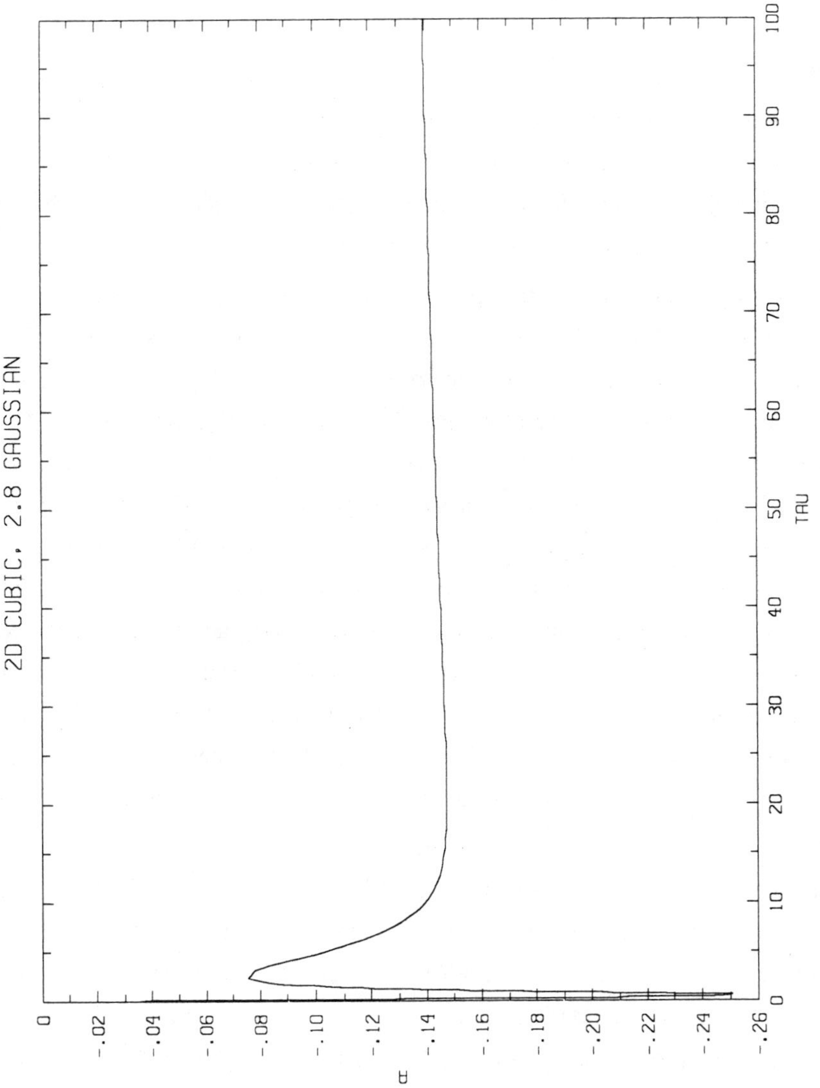
2D CUBIC, 2.8 GAUSSIAN
TAU
0
10
20
30
40
50
60
70
80
90
100
α
0
-.02
-.04
-.06
-.08
-.10
-.12
-.14
-.16
-.18
-.20
-.22
-.24
-.26

Figure 4c

One might conjecture that in all the other cases one is seeing convergence to a non-zero universal α value of approximately -0.177 as seen in the G2.77 case. Runs were attempted with G2.76 initial data and seemed to contradict this, giving a substantially smaller limit: $\alpha \approx -0.11$ However those runs have not yet been done without significant oscillations during the early stages of the run.

Profiles

As mentioned in section 9 there are theoretical arguments against $a(\tau)$ going to a non-zero limit related to the infinite L^2 norm of the Q profiles. This can be further studied by looking at convergence of profiles to a limit and attempting to fit both Q and R profiles to the final solution profile.

In the asymptotic form of the spatial profiles we shall again focus mostly on one choice of initial data, G4, with some comparisons to other cases. The common form of the final profiles for different initial data is illustrated in figures (5a,b) which compare the amplitude profiles at $\tau = 100$ for G4 and L4. The similarity is indeed far better than between profiles for the G4 run at times $\tau=90$ and 100, or even for times 100 and 200 when a run is done to the latter time. (this run is not tabulated here as accuracy was compromised somewhat to reach $\tau = 200$). This indicates that the "τ orbits in ξ space" for these two runs are not only converging to the same limit, but are moving along nearby paths to that limit. In other words, the agreement of the limiting behaviour at $\tau = \infty$ for singular solutions appears to be of order one in time rather than just order zero as previously conjectured. This will be discussed further below, where a proposed universal curve in function space leading to the universal profile is described.

Figures (6a-c) show the amplitude profiles at $\tau=50,60,\ldots,100$ on three ξ intervals. Convergence is again clear, though not as dramatic as in the supercritical case after $\tau = 10$. This convergence has been noted elsewhere, along with the observation that the limiting profile seems to be the ground state R_0 after a suitable similarity rescaling. However the possibility of a Q profile fit should also be studied. Indeed one feature of the profile graphs immediately suggests the Q profiles: they appear to have decay that is approximately $O(1/\xi)$ out to a large ξ value (after which the decay becomes more rapid). The width of this

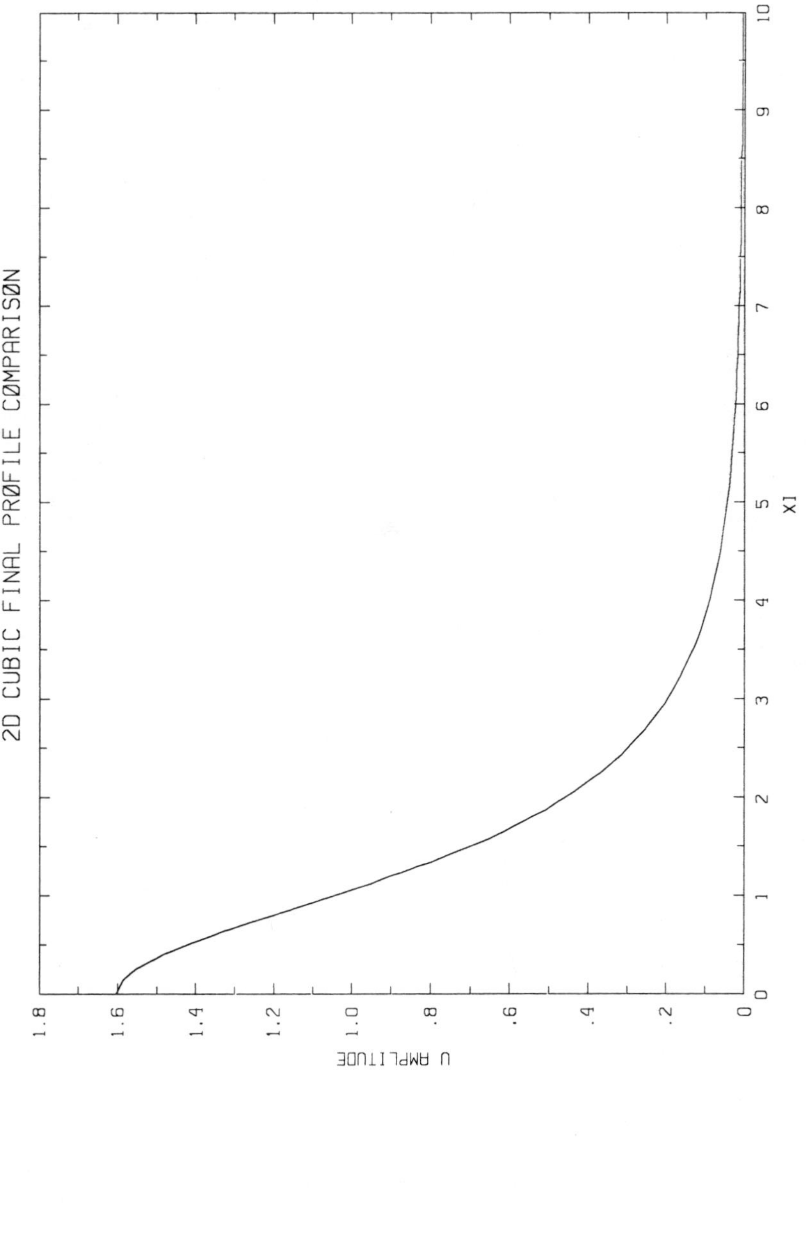
2D CUBIC FINAL PRØFILE CØMPARISØN
U AMPLITUDE
1.8
1.6
1.4
1.2
1.0
.8
.6
.4
.2
0
XI
0
1
2
3
4
5
6
7
8
9
10

Figure 5a

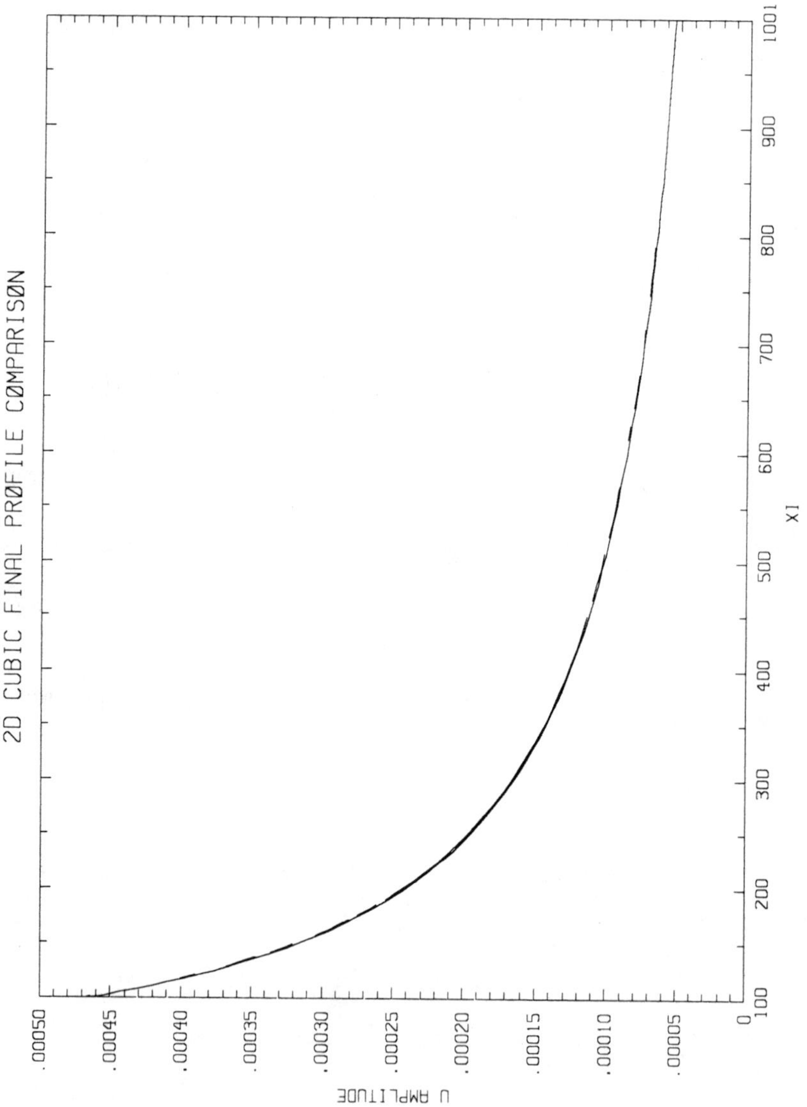
2D CUBIC FINAL PRØFILE CØMPARISØN
U AMPLITUDE
.00050
.00045
.00040
.00035
.00030
.00025
.00020
.00015
.00010
.00005
0
XI
100
200
300
400
500
600
700
800
900
1001

Figure 5b

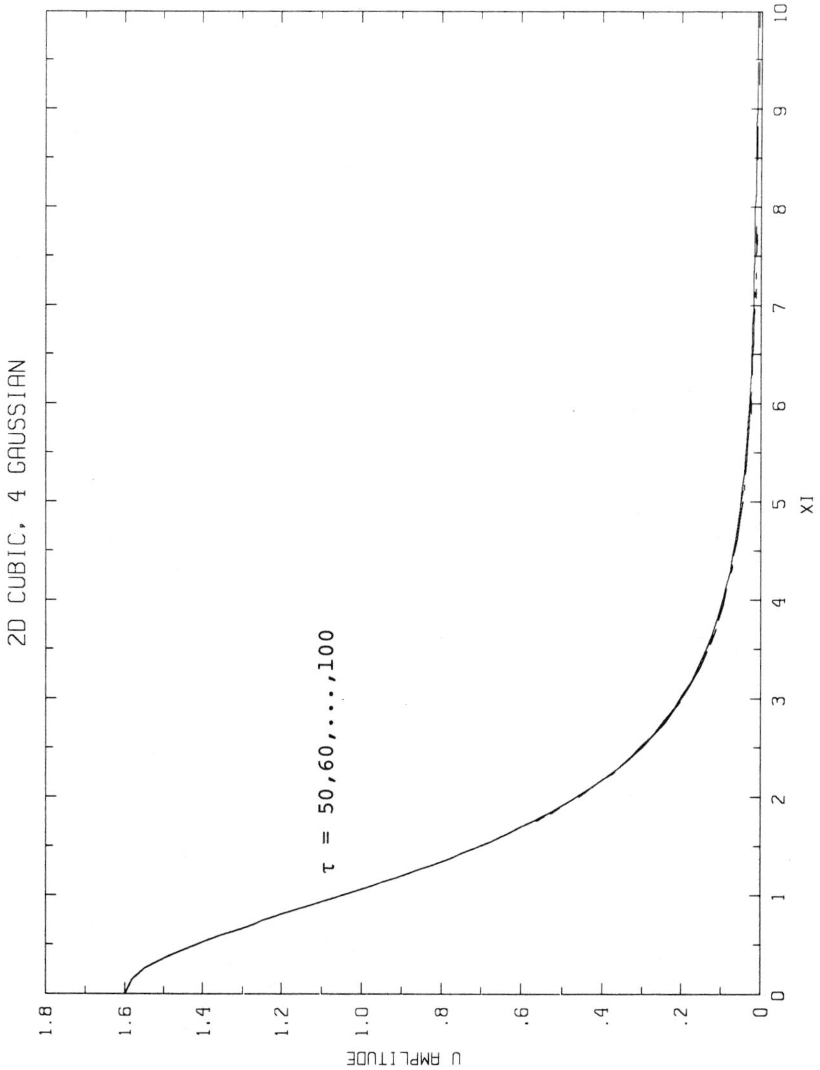
2D CUBIC, 4 GAUSSIAN
τ = 50,60,...,100
U AMPLITUDE
XI
0
.2
.4
.6
.8
1.0
1.2
1.4
1.6
1.8
0
1
2
3
4
5
6
7
8
9
10

Figure 6a

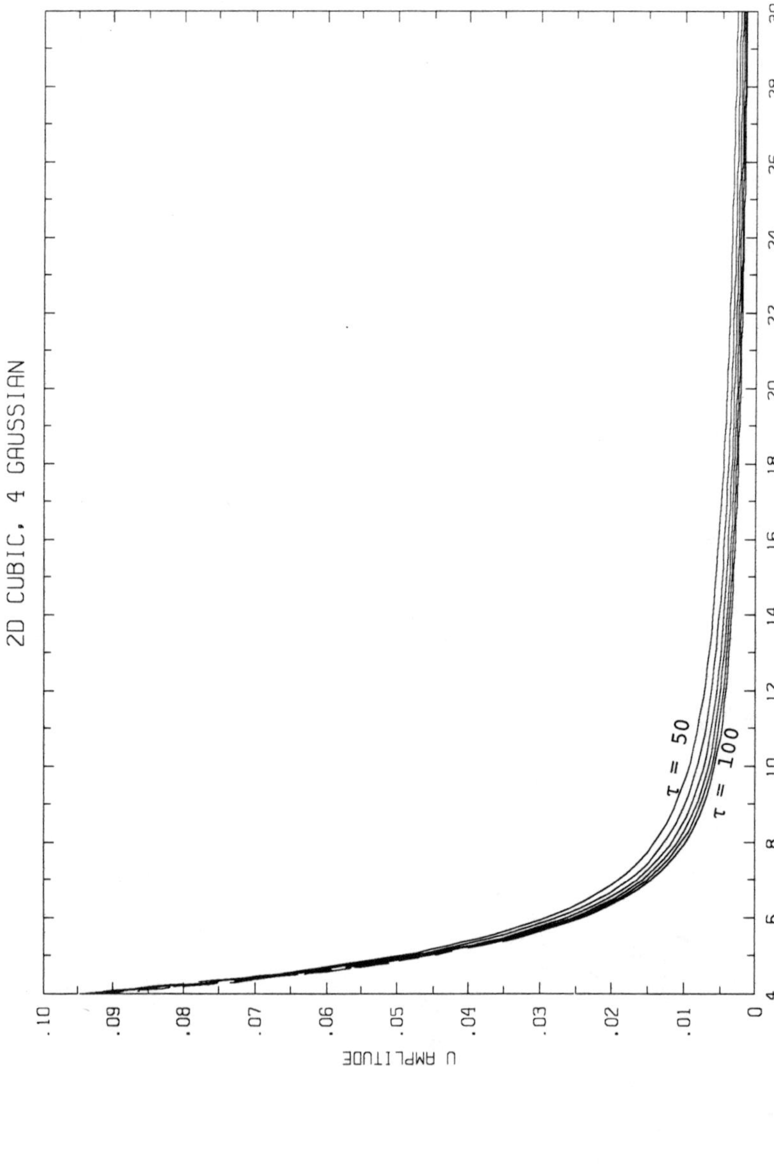
2D CUBIC, 4 GAUSSIAN
τ = 50
τ = 100
XI
U AMPLITUDE
4
6
8
10
12
14
16
18
20
22
24
26
28
30
0
.01
.02
.03
.04
.05
.06
.07
.08
.09
.10

Figure 6b

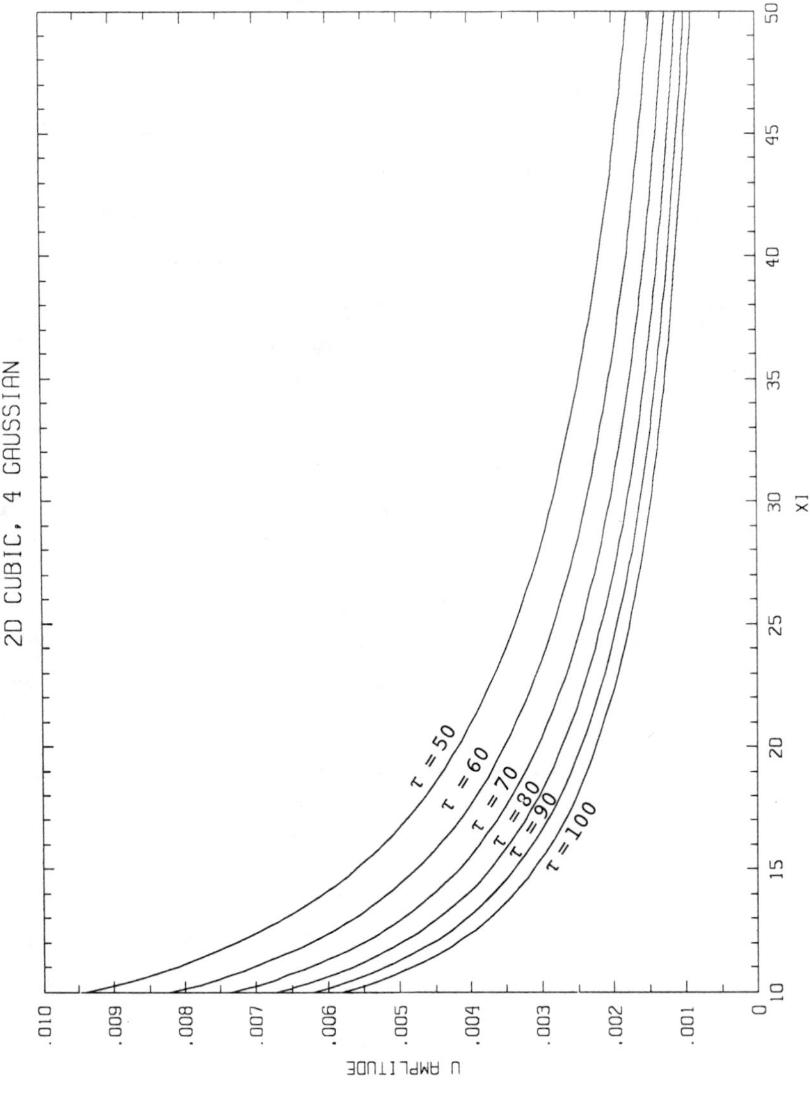
2D CUBIC, 4 GAUSSIAN
U AMPLITUDE
X1
τ = 50
τ = 60
τ = 70
τ = 80
τ = 90
τ = 100
.010
.009
.008
.007
.006
.005
.004
.003
.002
.001
0
10
15
20
25
30
35
40
45
50

Figure 6c

region increases with time so as to extend roughly out to $r = O(1)$. This resembles the behaviour of the Q profiles without the oscillatory factor (see section 9) more than the exponentially decaying R_0. On the other this slowly decaying part of the profile is also decaying as τ increases, increasing the size of a more rapidly decaying region nearer to the origin, so that eventual convergence to R_0 is not at all implausible.

The most obvious way to proceed is to fit a Q profile as was done previously, taking the parameters from the $\tau = 100$ data, and to do the same for R (setting $a_\infty = 0$ in effect). This done in figure 7(a). This comparison favours the Q profile distinctly: it is indistinguishable from the solution profile until about $\xi = 5.5$. However this is in a sense unfair because the exponential decay of R_0 is not generic: the R profiles in general oscillate around the constant solutions $R = \pm\sqrt{c}$ for large ξ. The Q profiles on the other hand always have the same qualitative behaviour, but with varying amplitudes for the oscillations. Since the fit of the R profile breaks down when it diverges from its decreasing trend towards the non-zero constant solution, it would be better to attempt to fit a rescaling of the ground state R_0, which empirically has $\mu_\infty = 2.2052$ for $c_\infty = 1$. This is done in figure (7b) where the rescaling using the c value from the final data and the corresponding initial value $\mu_\infty = 1.6378$. It is still not as good, but a slightly different rescaling might improve this and there is a common trend with the Q and R fits: they break down when the curve has its first zero or turning point.

It is worth enquiring whether the Q profile that we have just fitted to is in any way special. Given that $|a|$ is still decreasing it is also natural to wonder if a is converging to a smaller value that gives a distinguished Q profile. Both things are true. The Q equation was solved with the initial value and c_∞ fixed at those from the final data and a_∞ varied up and down. The result is that at for most other values the amplitude of the oscillations is substantially larger, but for small decreases in a_∞ the oscillations get smaller.

Figures (8a,b) show a remarkable relationship between this new Q and the final profile: the final data profile matches the envelope of the oscillations in Q. Thus there is a (μ, α) pair that is distinguished from other nearby pairs *with the same μ value*. However it is not unique in the manner of the pair found in the

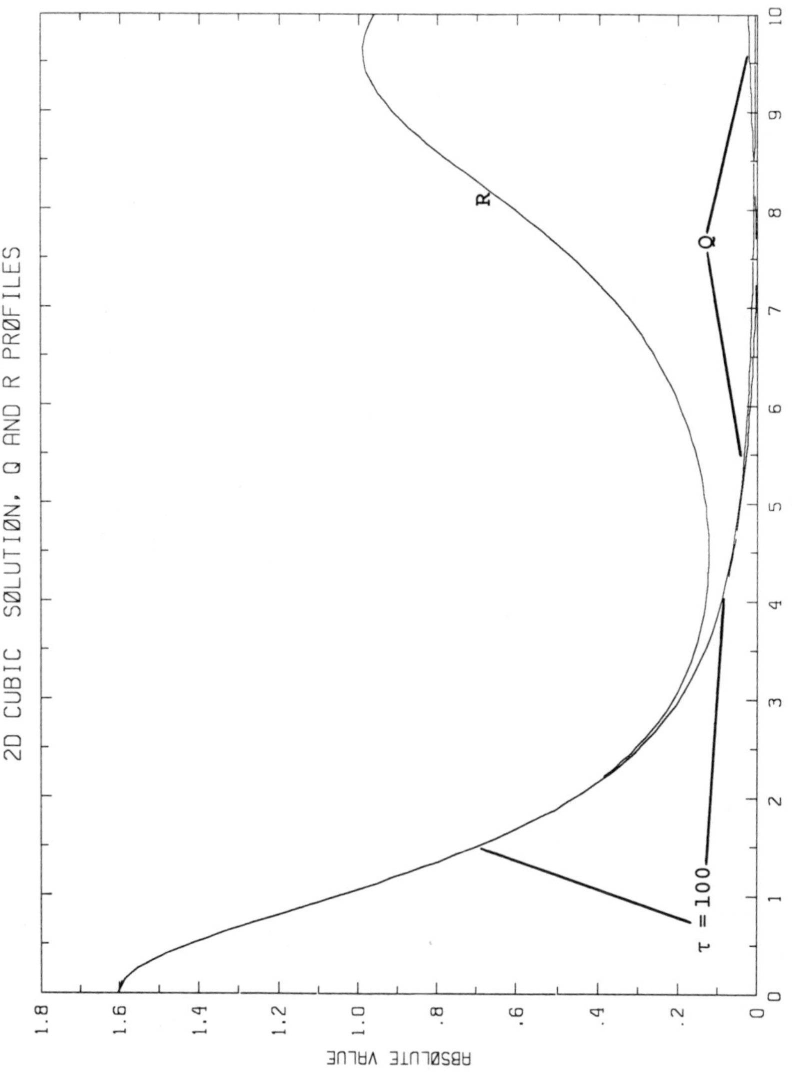

2D CUBIC SØLUTIØN, Q AND R PRØFILES
ABSØLUTE VALUE
XI
R
Q
τ = 100
0
.2
.4
.6
.8
1.0
1.2
1.4
1.6
1.8
0
1
2
3
4
5
6
7
8
9
10

Figure 7a

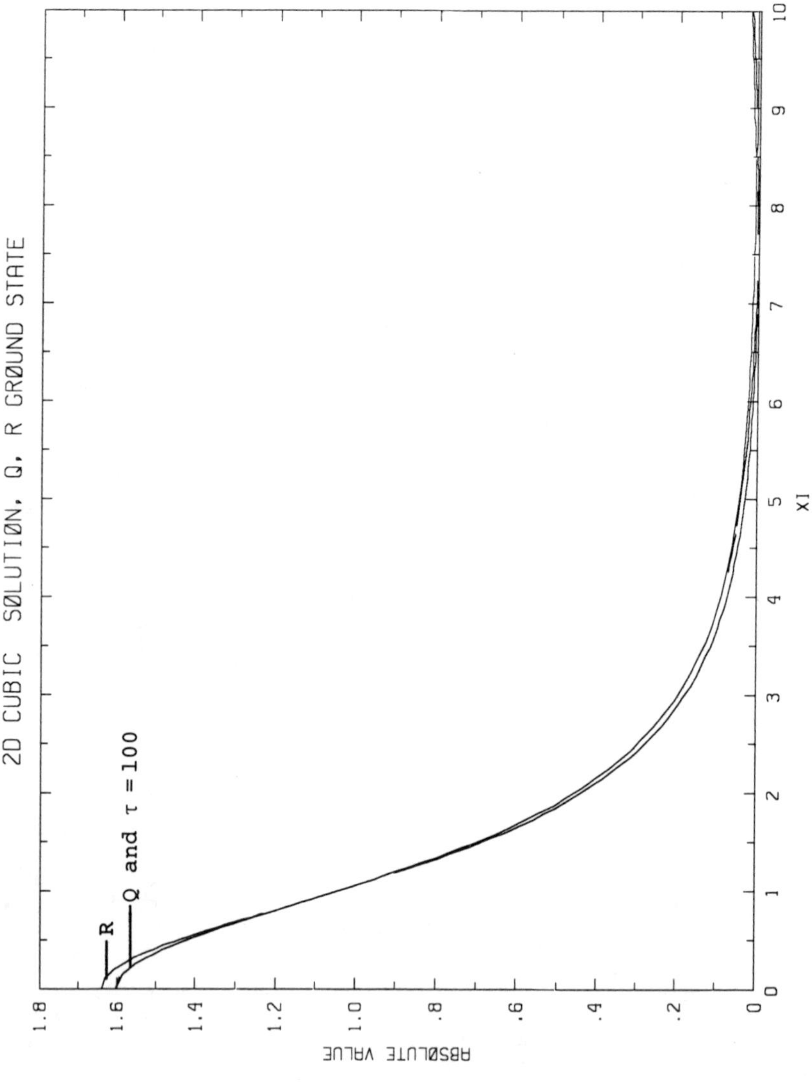

2D CUBIC SØLUTIØN, Q, R GRØUND STATE
R
Q and τ = 100
ABSØLUTE VALUE
XI
0
.2
.4
.6
.8
1.0
1.2
1.4
1.6
1.8
0
1
2
3
4
5
6
7
8
9
10

Figure 7b

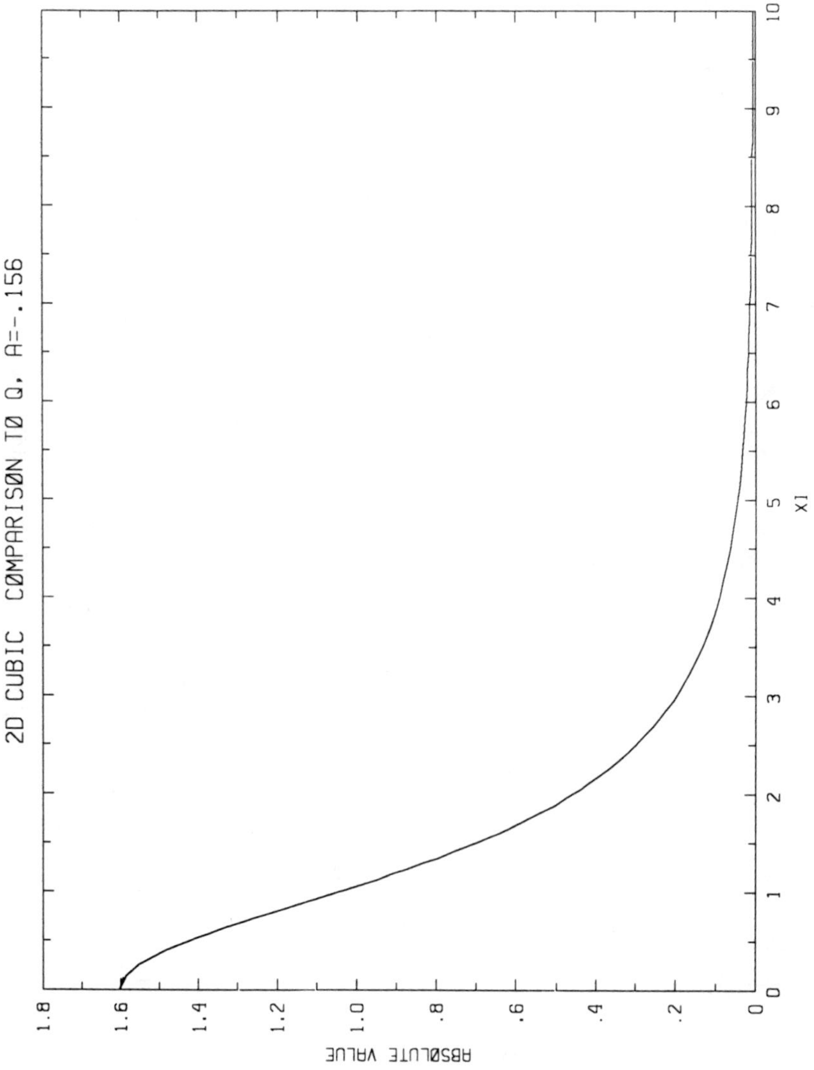
2D CUBIC COMPARISØN TØ Q, A=-.156
ABSØLUTE VALUE
XI
1.8
1.6
1.4
1.2
1.0
.8
.6
.4
.2
0
0
1
2
3
4
5
6
7
8
9
10

Figure 8a

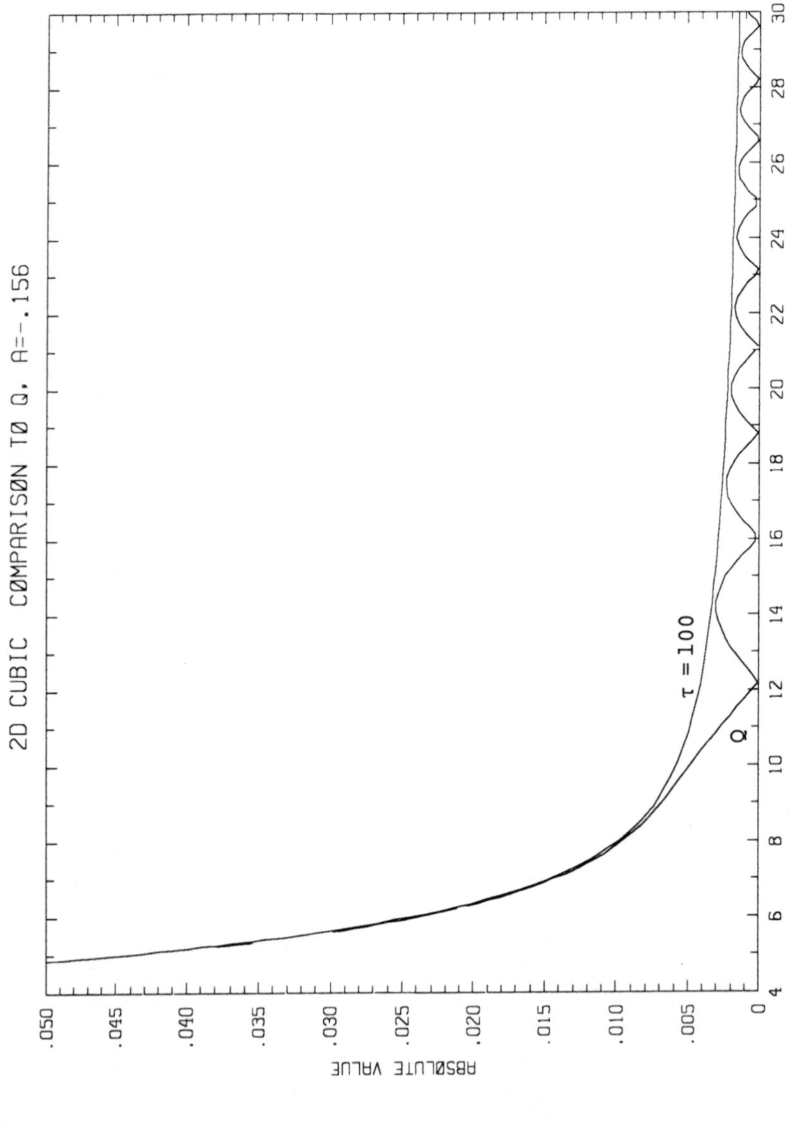
2D CUBIC COMPARISØN TØ Q, A=-.156
τ = 100
Q
ABSØLUTE VALUE
X1
.050
.045
.040
.035
.030
.025
.020
.015
.010
.005
0
4
6
8
10
12
14
16
18
20
22
24
26
28
30

Figure 8b

supercritical case. For the G2.8 and G2.77 cases have final α values that are smaller and going away from this value, and for the former, the Q profile fit to its final $|u|$, c and a values again has atypically small oscillations. This suggests that for the μ value for this run there is also a distinct "good" α value. They also have final μ values that are significantly closer to the value arising for the R_0 ground state, as shown in table 3. The trivial values for R_0 are included to support the suggestion that the other values lie near a common curve passing through it. The $\tau = 200$ values for G4 are from a second run in which the grid was expanded by setting $K = 26$ and which was continued until well after the boundary condition imposition at infinity had become meaningless. This gave $a(200) = -0.142175$ and $|u(200,0)| = 1.606164$. The former is beyond the

Table 3

Solution	$\mu_\infty - \mu$	α
L4 $\tau = 100$	0.0466	-0.28833
G4 $\tau = 100$	0.0465	-0.28798
G4 $\tau = 200$	0.0367	-0.25915
G2.8 $\tau = 100$	0.0353	-0.25458
G2.77 $\tau = 100$	0.0158	-0.1775
R_0	0	0

above distinguished a value and in fact the pair is much closer to the final values for the G2.8 run than to the earlier values with G4 initial data.

This suggests that there is an "oscillation minimizing" α value for each of a range of μ values near the above μ_∞, and that the near fit of the solution amplitude to such profiles moves along a continuum of them in the (μ,α) parameter space to R_0, which is clearly a particular case of such oscillation minimizers. This would reconcile the problem of the infinite L^2 norm of the Q profiles, since the approximate fit to any one of them is at finite τ and therefore only good (percentage wise) over a finite ξ interval. To state this conjecture more concisely:

> There is for each μ less than but near to $\mu_\infty, \approx 2.2052$, a value $\alpha(\mu) < 0$ such that the corresponding Q profile has minimal amplitude oscillations in some sense. This function is continuous with $\alpha(\mu_\infty) = 0$.
>
> For blowup solutions of the rescaled system, the parameterized curve $(\mu(\tau),\alpha(\tau))$ approaches $(\mu_\infty,0)$, asymptotically to the above "minimal

oscillation curve". Further the solution profiles, when rescaled to normalize c to 1, become increasingly close to the corresponding $Q(.,\mu,1,\alpha(\mu))$.

In the light of earlier comments on the apparent limiting behaviour of c, it is also reasonable to further conjecture that

If the rescaling is done with

$$G(u(\tau,.)) = G_0 = G(R_0(.)), = \|R_0\|_2^2$$

$c(\tau) \to 1$ and so no rescaling is needed in the above profile convergence.

References

[AK] Ablowitz, M. and Segur, H., *On the evolution of packets of water waves,* Journal of Fluid Mech. **92**, 691-715, 1979

[ASK] Akhmanov, S. A., Sukharukov, A. P. and Kokhlov, R. V., *Self-focusing and self-trapping of intense light beams in a nonlinear medium,* Translated in Soviet Physics JETP **23**, 1025-1033, 1966

[BC] Bluman, G. W. and Cole, J. D., *Similarity methods for differential equations,* Springer, N.Y., 1974

[BZS] Budneva, O. B., Zakharov, V. E. and Synakh, V. S., *Certain models for wave collapse,* Translated in Soviet Journal of Plasma Physics **1**, 335-338, 1975

[CGT] Chiao, R. Y., Garmire, E. and Townes, C. H., *Self-trapping of optical beams,* Phys. Rev. Lett. **13**, 479-482, 1964

[C] Coffman, C. V., *Uniqueness of the ground state for $\Delta u - u + u^3 = 0$ and a variational characterization of other solutions,* Archive for Rational Mechanics and Analysis **46**, 81-95, 1972

[ES] Elliot, C. J. and Suydam, B. R., *Self-focusing phenomena in air-glass laser structures,* IEEE Journal of Quantum Electronics **11**, 863-866, 1975

[GV] Ginibre, J. and Velo, G., *On a class of nonlinear Schrödinger equations. I. The Cauchy problem, general case,* Journal of Functional Analysis **32**, 1-32, 1979

[Gl] Glassey, R. T., *On the blowing up of the solutions to the Cauchy problem for nonlinear Schrödinger equations,* Journal of Mathematical Physics **18**, no. 9, 1974-1977, 1977

[Go] Goldman, M. V., *Strong turbulence of plasma waves,* Rev. Mod. Phys. **56**, 709-735, 1984

[GRH] Goldman, M. V., Rypdal, K. and Hafizi, B., *Dimensionality and dissipation in Langmuir collapse,* Physics of Fluids **23**, 945-955, 1980

[K] Kusnetsov, E. A., *The collapse of electro-magnetic waves in a plasma,* Zh. Eksp. Teor. Fiz. **66**, 2037-2047, 1974,
Translated in Soviet Physics JETP **39**, 1003-1007, 1974

[L] Lees, M., *A linear three-level difference scheme for quasilinear parabolic equations,* Math. Comp. **20**, 516-522, 1966

[MP] McLaughlin, D. W. and Papanicolaou, G. C., unpublished notes, October 1982

[MPSS] McLaughlin, D. W., Papanicolaou, G. C., Sulem, C. and Sulem P. L., *The focusing singularity of the cubic Schrödinger equation,* Physical Review A, to appear

[MS] McLeod, K. and Serrin, J., *Uniqueness of solutions of semilinear Poisson equations,* Proc. Nat. Acad. Sci. U.S.A. **78**, no. 11, 6592-6595, 1981

[N] Newell, A. C., *Soliton perturbations and nonlinear focussing,* in *Solitons and condensed matter physics,* Bishop, A. R. and Schneider, T., editors, Springer Verlag, Berlin, 52-67, 1978

[O] Ovsiannikov, L. V., *Group properties of differential equations,* Bluman, G. W. translator, Caltech 1975

[PMW] Papanicolaou, G., McLaughlin, D. and Weinstein, M. *Focusing singularity for the nonlinear Schrödinger equation,* Lecture Notes in Num. Appl. Anal. **5**, 253-257, 1982

[Sul] Sulem, C., *Etude théorique et numérique de quelques problèmes mathématiques en hydrodynamique: Regularité et apparition de singularités,* Thesis, l'Université Paris-Nord, 1982

[SSF] Sulem, C., Sulem, P. L. and Frisch, H., *Tracing complex singularities with spectral methods,* Journal of Comp. Phys. **50**, no. 1, 138-161, 1983

[SSP] Sulem, P. L., Sulem, C. and Patera, A., *Numerical simulation of singular solutions to the two-dimensional cubic Schrödinger equation,* Comm. in Pure and Applied Math. **37**, 755-778, 1984

[Sud] Suydam, B. R., IEEE Journal of Quantum Electronics **11**, 225, 1975

[T1] Talanov, V. I., *Self-focusing of wave beams in nonlinear media,* Translated in JETP Lett. **2**, 138, 1965

[T2] Talanov, V. I., *Self-modelling wave beams in a nonlinear dielectric,* Izvestia VUZ. Radiofizika 9, no. 2, 410-412, 1966
Translated in Soviet Radiophysics **9**, 260-261, 1967

[VPT] Vlasov, S. N., Piskunova, L. V. and Talanov, V. I., *Structure of the field near a singularity arising from self-focusing in a cubically nonlinear medium,* Zh. Eksp. Teor. Fiz. **75**, 1602-1609, 1978
Translated in Soviet Physics JETP **48**, 808-812, 1978

[We1] Weinstein, M. I., *Self-focusing and modulation analysis for nonlinear Schrödinger equations,* Ph. D. Thesis, New York University, 1982

[We2] Weinstein, M. I., *Nonlinear Schrödinger equations and sharp interpolation estimates,* Comm. in Mathematical Physics **87**, 567-576, 1983

[We3] Weinstein, M. I., *Modulation stability of ground states of nonlinear Schrödinger equations,* SIAM Journal of Mathematical Analysis **16**, no.3, 472-491, 1985

[Wo] Wood, D., *The self-focussing singularity in the nonlinear Schrödinger equation* Studies in Applied Mathematics **71**, 103-115, 1984

[Z1] Zakharov, V. E., *Collapse of Langmuir waves,* Translated in Soviet Physics JETP **35**, 908-922, 1972

[Z2] Zakharov, V. E., *Collapse of self-focussing of Langmuir waves* in *Handbook of Plasma Physics vol. 2,* Rosenbluth, M. N. and Sagdeev, R. Z. editors, Elsevier, 1984

[ZSh] Zakharov, V. E. and Shabat, A. B., *Exact theory of two-dimensional self-focusing and one-dimensional self-modulation of waves in nonlinear media,* Zh. Eksp. Teor. Fiz. 61, 118-134, 1971
Translated in Soviet Physics JETP **34**, 62-69, 1972

[ZSy] Zakharov, V. E. and Synakh, V. S., *The nature of the self-focusing singularity,* Zh. Eksp. Teor. Fiz. **68**, 940-947, 1975
Translated in Soviet Physics JETP **41**, 465-468, 1976

Acknowledgement

This work was supported by the Air Force Office of Scientific Research under Grant AFOSR 85-0017.

Department of Mathematical Sciences
Rensselaer Polytechnic Institute
Troy, N. Y., 12180

Courant Institute of Mathematical Sciences
251 Mercer Street
N. Y., N. Y., 10012

Department of Mathematics and Computer Science
Ben Gurion University of the Negev
P. O. Box 653 Beer-Sheva
84 105, Israel

Department of Mathematics
Tel Aviv University
Ramat Aviv, Israel

A Probabilistic Approach to Finding Estimates for the Heat Kernel Associated with a Hörmander Form Operator

Daniel W. Stroock

0. Introduction:

This note is a report on some recent work by S. Kusuoka and the present author [2].

Let $V_1,\ldots,V_d \in C_b^\infty(R^N;R^N)$ and, thinking of $V \in C_b^\infty(R^N;R^N)$ as the directional derivative operator $\sum_{i=1}^N V_i(x)\partial_{x_i}$, define $V^{(\alpha)}$, $\alpha \in \mathcal{A} \equiv \{\phi\} \cup \bigcup_{\ell=1}^N (\{1,\ldots,d\})^\ell$, inductively by

$$V^{(\alpha)} = \begin{cases} V \text{ if } \alpha = \phi \\ [V_k, V^{(\alpha')}] \text{ if } \alpha = (\alpha',k)' \end{cases}$$

where $[V,W]$ denotes the Lie product (i.e. commutator) of V and W. Assume that

$$\text{(H)} \qquad \sum_{k=1}^d \sum_{|\alpha|\le \ell_0} (V_k^{(\alpha)},\eta)^2_{R^N} \ge \epsilon|\eta|^2,\ \eta \in R^N,$$

for some $\ell_0 \ge 0$ and $\epsilon > 0$. (Given $\alpha \in \mathcal{A}$,

$$|\alpha| \equiv \begin{cases} 0 \text{ if } \alpha = \phi \\ \ell \text{ if } \alpha \in (\{1,\ldots,d\})^\ell \end{cases}.)$$

Next, define H to be the Hilbert space of $h \in C([0,\infty);R^d)$ such that $h(t) = \int_0^t h'(s)ds$ for some $h' \in L^2([0,\infty);R^d)$ and set $\| h \|_H \equiv \| h' \|_{L^2([0,\infty);R^d)}$. Given $h \in H$, define $(t,x) \in [0,\infty)\times R^N \longrightarrow X(t,x;h) \in R^N$ by

DIRECTIONS IN PARTIAL
DIFFERENTIAL EQUATIONS

ISBN 0-12-195255-X

$$(0.1)\qquad X(T,x;h) = x + \sum_{k=1}^{d} \int_0^T V_k(X(t,x;h))h'(t)dt,\ T \geq 0.$$

Then, because of (H), elementary control theory allows one to check that for all $x,y \in R^N$ there exists an $h \in H$ such that $X(1,x;h) = y$. Moreover, if

$$d(x,y) \equiv \inf\{\| h \|_H : X(1,x;h) = y\},$$

then d is a complete metric on R^N which is compatible with the Euclidean topology. For $x \in R^N$ and $r > 0$, define $B_d(x,r)$ to be $\{y \in R^N: d(x,y) < r\}$.

Finally, let $a_1,\ldots,a_d \in C_b^\infty(R^N)$ (when no range is specified, the range is assumed to be R^1) and set

$$V_0 = \sum_{k=1}^{d} a_k V_k$$

and define

$$L = 1/2 \sum_{k=1}^{d} V_k^2 + V_0.$$

(V^2 denotes the second iterate of the directional derivative operator V.) Our main result is contained in the following statement.

(0.2) Theorem: There exists a unique positive $p \in C^\infty((0,\infty)xR^N xR^N)$ such that $\int p(t,x,y)dy = 1$ for all $(t,x) \in (0,\infty)xR^N$ and, for all $f \in C_b(R^N)$, the function $u_f(t,x) = \int f(y)p(t,x,y)dy$ is the unique $u \in C^\infty((0,\infty)xR^N) \cap C_b(R^N)$ satisfying

$$\partial_t u = Lu$$

$$\lim_{t \searrow 0} u(t,.) = f.$$

Moreover, there exist a $\lambda \in (0,1)$ such that

$$(0.3)\qquad \frac{\lambda}{|B_d(x,r)|}\exp(-d(x,y)^2/\lambda t) \le p(t,x,y) \le \frac{1}{\lambda|B_d(x,r)|}\exp(-\lambda d(x,y)^2/t)$$

for all $(t,x,y) \in (0,1]\times R^N\times R^N$. (If Γ is a measurable subset of R^N, then $|\Gamma|$ denotes its Lebesgue volume.)

(0.4) Remark: Theorem (0.2) is a variation on a theorem proved originally by A. Sanchez [7] and improved recently by him and D. Jerison [5]. The approach taken in both those articles is based on ideas introduced in L. Rothschild and E. Stein [6] and developed further in C. Fefferman and D. Phong [1]. Our proof, which is outlined below, can be viewed as a probabilistic interpretation of the ideas in [6].

As a quite easy corollary of Theorem (0.2), we obtain the following Moser-type version of the Harnack inequality for the parabolic operator $\partial_t + L$.

(0.5) Corollary: Let $(T_0,x^0) \in R^1\times R^N$ and $r \in (0,1]$ be given. Assume that u is a positive solution to $\partial_t u + Lu = 0$ in $(T_0,T_0+4r^2)\times B_d(x^0,2r)$ Then there exists an M (which is independent of T_0, x^0, and r) such that

$$u(t,x) \le u(s,y)\exp(M(1 + d(x,y)^2/t))$$

for all $(s,x), (t,y) \in [T_0,T_0+r^2]\times B_d(x^0,r)$.

(0.6) Remark: At least when L is formally self-adjoint and u is independent of t, the preceding corollary has been obtained in [4] by D. Jerison. Our proof has the considerable advantage over his that it relies only on the estimate (0.3) and requires very little additional work. In particular, because his approach is modeled after the famous one used by Moser, Jerison needs to first derive the Poincaré inequality corresponding to L.

1. Outline of the Proof:

Because the general case can be easily derived from the one when $V_0 = 0$, we will assume from the outset that $L = 1/2 \sum_{k=1}^{d} V_k^2$.

Our first step is to construct the diffusion associated with the operator L. To this end, let $(\beta(t), \mathcal{F}_t, P)$ be a standard d-dimensional Brownian motion and, for $x \in R^N$, define $X(.,x)$ by the stochastic integral equation

$$(1.1) \qquad X(T,x) = x + \sum_{k=1}^{d} \int_0^T V_k(X(t,x)) \circ d\beta_k(t), \quad T \geq 0.$$

(The notation "$\circ d\beta_k(t)$" means that the stochastic integrals in (1.1) are taken in the sense of Stratonovich.) Then, as a familiar consequence of Itô's theory, $(t,x) \in (0,\infty) x R^N \longrightarrow P(t,x,.) \equiv P \circ X(t,x)^{-1}$ is a continuous transition probability function and

$$(1.2) \qquad \int \varphi(y) P(T,x,dy) - \varphi(x) = \int_0^T dt \int L\varphi(y) P(t,x,dy)$$

for all $\varphi \in C_b^2(R^N)$. From (1.2) it is clear that

$$\partial_t P(t,x,.) = (L_y)^* P(t,x,.)$$

in the sense of distributions. Hence, by Hörmander's theorem, $P(t,x,dy) = p(t,x,y)dy$ where $(t,y) \in (0,\infty) x R^N \longrightarrow p(t,x,y)$ is smooth. Actually, a slightly refined version of Hörmander's theorem in [3] says that p is a positive, smooth function on $(0,\infty) x R^N x R^N$ and that, for $n \geq 0$, there exist $C_n < \infty$ and ν_n, $\mu_n > 0$ such that

$$(1.3) \qquad |\partial_x^\alpha \partial_y^\beta p(t,x,y)| \leq C_n / t^{\nu_n} \operatorname{esp}(-\mu_n |y - x|^2 / t)$$

for all $|\alpha| + |\beta| \leq n$ and $(t,x,y) \in (0,1] x R^N x R^N$.

Clearly $p(t,x,y)$ is the function discussed in Theorem (0.2) and all that remains to check is that the estimate (0.3) holds. However, before describing our method, it might be useful to point out that (0.3) is "obvious" from (1.1). Indeed:

$$\text{"}p(t,x,y) = P(X(t,x) = y) = \text{const.}\int \exp(-1/2\int_0^t |h'(s)|^2 ds)\delta h\text{"}$$

where the last expression is the famous (non-sense) "flat integral" representation of Wiener measure in terms of the (non-existent) Lebesgue measure on H. In particular, it seems clear that

$$\log(p(t,x,y)) \approx -\inf\{\| h \|_H : X(t,x;h) = y\} = -d(x,y)^2/t.$$

Recalling that $\int p(t,x,y)dy = 1$ and realizing from the preceding that $p(t,x,.)$ is almost supported on $B_d(x,t^{1/2})$, (0.3) becomes plausible. Actually, even though the preceding line of reasoning cannot bear too close scrutiny, it is the intuitive basis for our proof of (0.3). In fact, what we do is carry out the preceding argument in a finite dimensional setting, where it is easier to make rigorous.

The first step is to replace $X(t,x)$, which is a highly complicated function of $\beta(.)$, by its ℓth Taylor polynomial:

$$X_\ell(t,x) \equiv x + \sum_{\alpha\in A_\ell} \beta^{(\alpha)}(t)V_{(\alpha)}(x)$$

where $A_\ell \equiv \{\alpha \in A: 1 \le |\alpha| \le \ell\}$,

$$\beta^{(\alpha)}(t) \equiv \begin{cases} \beta_k(t) \text{ if } \alpha = (k) \\ \int_0^t \beta^{(\alpha')}(s)\circ d\beta_k(s) \text{ if } \alpha = (\alpha',k), \end{cases}$$

and

$$V_{(\alpha)}(x) \equiv \begin{cases} V_k(x) \text{ if } \alpha = (k) \\ \sum_{i=1}^N V_k^i(x)\partial_{x_i} V_{(\alpha)}(x) \text{ if } \alpha = (\alpha',k). \end{cases}$$

It is easy to see from (1.1) that

$$(1.4) \qquad \sup_{x\in R^N} E[|X(t,x) - X_\ell(t,x)|] \le Ct^{(\ell+1)/2},\ t \in [0,1].$$

In view of (1.4), it is believable that estimates on the distribution of $X_\ell(t,x)$ should be transferable as estimates on the distribution of $X(t,x)$. Thus, we turn our attention to $X_\ell(t,x)$.

The advantage which $X_\ell(t,x)$ has over $X(t.x)$ is that it is a much simpler function of $\beta(.)$; indeed, it is a linear function of the quantities $\beta^{(\alpha)}(t)$, $\alpha \in \mathcal{A}_\ell$. In particular, $PoX(t,x)^{-1}$ is the image of $Po(\{\beta^{(\alpha)}(t): \alpha \in \mathcal{A}_\ell\})^{-1}$ under a very simple mapping of $R^{\mathcal{A}_\ell}$ into R^N. Thus, it is the replacement of $X(t,x)$ by $X_\ell(t,x)$ which accomplishes the move, alluded to above, from an infinite to a finite dimensional setting. However, before we can carry out our program, it is necessary for us to learn quite a lot about the distribution of the $\beta^{(\alpha)}(t)$'s.

The key to studying $Po(\{\beta^{(\alpha)}(t): \alpha \in \mathcal{A}_\ell\})^{-1}$ is the observation that $\{\beta^{(\alpha)}(.): \alpha \in \mathcal{A}_\ell\}$ is a "Gaussian process" on a certain nilpotent Lie group $G \subseteq R^{\mathcal{A}_\ell}$. To see this, define the vector fields W_k, $1 \le k \le d$, on $R^{\mathcal{A}_\ell}$ by

$$W_k \equiv \sum_{\alpha\in\mathcal{A}_\ell} z_\alpha \partial_{z_{(\alpha,k)}},$$

($z \in R^{\mathcal{A}_\ell} \longrightarrow z_\alpha$, $\alpha \in \mathcal{A}_\ell$, are the standard coordinate maps) and define $Z(.,z)$, $z \in R^{\mathcal{A}_\ell}$, so that

$$Z(T,z) = z + \sum_{k=1}^{d} \int_0^T W_k(Z(t,z)) od\beta_k(t), \quad T \ge 0.$$

It is then easy to check that $Z(.,0) = \{\beta^{(\alpha)}(.): \alpha \in \mathcal{A}_\ell\}$. Next, set $\mathfrak{G} = Lie(W_1,\dots,W_d)$ and $G = \{\exp(W)(0): W \in \mathfrak{G}\}$; and note that, since $\mathfrak{G}$ is nilpotent, G can be made into a Lie group in which $\exp(W)(0)x\exp(W')(0) = \exp(W)(\exp(W')(0))$. Because the W_k's are obviously G-right invariant, we now see that $\{\beta^{(\alpha)}(t): \alpha \in A_\ell\}$ is the Gaussian process on G generated by $1/2 \sum_{k=1}^{d} W_k^2$; and because, by construction, $\mathrm{Lie}(W_1,\ldots,W_d)$ is the tangent space to G, Hörmander's theorem applies and says that $P\circ Z(t,z)^{-1}$ admits a smooth density $q(t,x,.)$ with respect to the Haar measure λ on G. Clearly, by right invariance, $q(t,z,\xi) = \rho(t,\xi z^{-1})$. In addition, since $\mathfrak{G}$ can be graded, G has a natural scale invariance which is reflected in ρ by the fact that $\rho(t,.)$ can be obtained from $\rho(1,.)$ by a rescaling. Using these observations, it is now quite easy to show that $q(t,z,\xi)$ satisfies an estimate of the sort in (0.3), only here with the metric d which is there replaced by an analogous metric defined in terms of the W_k's.

Once $P\circ(\{\beta^{(\alpha)}(t): \alpha \in A_\ell\})^{-1}$ is well understood, one can use it to play the same role in the analysis of $P\circ X_\ell(t,x)^{-1}$ as Wiener measure played in our heuristic analysis of $P\circ X(t,x)^{-1}$. In this way one obtains estimates of the form in (0.3) for $P\circ X_\ell(t,x)^{-1}$; and these can, in turn, be transferred to $P\circ X(t,x)^{-1}$ via (1.4).

REFERENCES

[1] C. Fefferman and D. Phong, "Subelliptic eigenvalue problems," Proc. Conf. on Harmonic Analysis in Honor of A. Zygmund, publ. by Wadsworth Math. Series, pp. 590-606 (1981).

[2] S. Kusuoka and D. Stroock, "Applications of the Malliavin Calculus, Part III," in preparation.

[3] S. Kusuoka and D. Stroock, "Applications of the Malliavin Calculus, Part II," J. Fac. Sci. Univ. Tokyo, Sect. IA, Math., 32, pp. 1-76(1985).

[4] D. Jerison, "The Poincare Inequality for Vector Fields Satisfying Hormander's Condition," preprint.

[5] D. Jerison and A. Sanchez-Calle, "Estimates for the Heat Kernel for a sum of Squares of Vector Fields," preprint.

[6] L. Rothschild and E. Stein, " Hypoelliptic Differential Operators and Nilpotent Groups," Acta. Math. 137, pp. 247-320, (1984).

During the period of this research, the author was partially supported by N.S.F. Grant D.M.S-8415211 and A.R.O. Grant DAAG29-84-K-0005.

Department of Mathematics
Massachusetts Institute of Technology
Cambridge, MA 02139

Discontinuities and Oscillations

Luc C. Tartar

The subject under study here was initiated more than a century ago as regards discontinuities - then it took some time for physicists to agree on what the problems were. One had to wait a little longer before mathematicians became interested (Courant and Friedrichs [1], 1948) and then became able to ask themselves precise questions about discontinuities (Lax [2], 1957). However, as pointed out by Dafermos ([3], 1983), despite considerable progress in recent years, most of the fundamental problems in the analytical theory remain unsolved.

As regards oscillations the situation is not as clear, first because the mathematical results (at least those I know and use) are recent and second because there may be disagreement with and among physicists about the relevance of these mathematical tools for describing physical reality.

Having found how some mathematical results, obtained in joint work with F. Murat, could give me a better understanding of some questions of Physics than what I learned from my Physics teachers (who happended to be all physicists), I was naturally led to try to understand more about Physics.

After some time I became convinced that a new mathematical tool was necessary and that only mathematicians (with a deep interest in Science) could develop it.

DIRECTIONS IN PARTIAL
DIFFERENTIAL EQUATIONS

ISBN 0-12-195255-X

My own attempts have been centered on the study of oscillations in nonlinear partial differential equations: hyperbolic, quasilinear or semilinear systems provided a good testing ground for these questions and I will outline some of these attempts later on.

There is a very common source of disagreement coming from the identification of Physics and physicists (or Mathematics and mathematicians). Physics has the goal of identifying the basic laws of Nature and physicists are the people who have been trained by learning the previous attempts to discover them (is it really a good training for discovering more? In some cases accepting what was "found" before has been a real handicap and light came only after rejecting part of the "authorized" explanations!). From the identification of the goal with the attempts, Physics is also considered to be the collection of questions studied by physicists, which is indeed a different definition which includes part of Technology and of Mathematics.

Mathematicians are trained by learning the previous attempts to understand Mathematics which, although I followed this training with reasonable success, I will not dare to define precisely. Being a mathematician interested in Science, I am naturally interested in Physics, which I interpret through my understanding of Mathematics and from what I learn about the conception the physicists have of it. It is my belief that what I do is useful for understanding Physics. The fact that I am not working in the authorized direction does not prove that I am wrong.

Because of difficulties I have had with some of my excolleagues, I can work only in a friendly atmosphere and Madison is one of the few places where I always felt secure.

It is a great pleasure for me to emphasize the fact that a large portion of my work has been done in Madison while visiting the Mathematics Research Center.

This was certainly due to the warm atmosphere of Madison and the support of all my friends, of whom I want to thank particularly John Nohel and Mike Crandall.

John Nohel arranged my first visit to the United States in April 1971: there was a similar Symposium here related to my subject in Lax's important paper on entropy conditions [35]. Later he arranged a one year visit, 1974-1975, where I held a joint appointment between the Department and the Mathematics Research Center.

This was a very profitable year where I started a fruitful collaboration with Mike Crandall (although we did not publish much of our results) and I simplified most of the results I had obtained with F. Murat on homogenization by understanding how to use our div-curl lemma. I came then for two months at the time of another Symposium in October 1977 and there I found the missing link for my method of using the div-curl lemma to prove convergence of approximate solutions for a scalar hyperbolic equation [46]. I spent also the summer of 1980 here and it was during that stay that I developed the first results on propagation and interaction of oscillations for the Carleman or the Broadwell models [56]. I also improved a method that I had found to study some special semilinear systems and obtained for them results of global existence and asymptotic behaviour [55].

I want to outline here the evolution of some of the ideas that I have developed and how, in order to understand discontinuities, I was led to study oscillations. This survey will not be exhaustive; it describes only my own attempts.

1. DISCONTINUITIES; PART ONE. ([1], 1948)

It was noticed by Challis ([4], 1848) that a formula obtained by Poisson ([5], 1808) giving a simple wave solution for flow in an isothermal gas could not always be solved; then Stokes ([6], 1848) proposed that discontinuities should be allowed and he found two discontinuity conditions from conservation of mass and momentum. Earnshaw ([7], 1860) did a more general case but it was Riemann ([8], 1860) who gave the general solution and introduced the so-called Riemann invariants, unfortunately working with an hypothesis of isentropy.

The understanding of what kind of discontinuities were physical came after the work of Rankine ([9], 1870), Hugoniot ([10], 1889) and Rayleigh ([11], 1910).

In order to summarize the results of this period with modern notation, let us consider the following Cauchy problem

$$u_t + f'(u)u_x = 0 \tag{1.1}$$

$$u(x,0) = v(x) \ . \tag{1.2}$$

Considering (1.1) as a partial differential equation of first order with a variable coefficient $f'(u(x,t))$ we introduce the characteristics given by the ordinary differential equation

$$\frac{dx(t)}{dt} = f'(u(x(t),t)) \tag{1.3}$$

and then notice that (1.3) implies

$$\frac{du(x(t),t)}{dt} = 0 \ . \tag{1.4}$$

The characteristics are then straight lines on which u is constant and this leads to the formula

On $x = x_0 + tf'(v(x_0))$, one has

$$u(x,t) = v(x_0) \ . \tag{1.5}$$

Then one points out that, except in the case where $f'(v(x))$ is nondecreasing, two different characteristics will meet and (1.5) cannot be used.

One can actually deduce from (1.5) the critical time after which it becomes ambiguous:

$$T_0 = \frac{-1}{\operatorname{Inf} f''(v(x))} \ . \tag{1.6}$$

The preceding computations were made under the assumption that the solution was smooth and it led to (1.5) which implies that there is no bound for derivatives of u as t goes to T_0.

We resolve this dilemma by accepting discontinuous solutions and rewrite (1.1) in conservation form

$$u_t + (f(u))_x = 0 \tag{1.7}$$

where we use derivatives in the sense of distributions and this implies a jump condition on a curve of discontinuity. If our solution has left and right limits u_- and u_+ on a smooth curve $x(t)$, its slope $s = x'(t)$ is given by

$$s = \frac{f(u_+) - f(u_-)}{u_+ - u_-} . \tag{1.8}$$

In the case where the unknown function U has p components we will restrict to systems written in conservation form:

$$U_t + (F(U))_x = 0 . \tag{1.9}$$

The jump condition becomes a system of p equations relating to the velocity of the discontinuity with the left and right limits U_+ and U_-:

$$F(U_+) - F(U_-) = s(U_+ - U_-) . \tag{1.10}$$

Although the analog of (1.10) for the system of gas dynamics was first derived by Stokes and then rediscovered by Riemann, these conditions are now named after Rankine and Hugoniot.

If we consider equation (1.7) with discontinuous data (the Riemann problem)

$$v(x) = u_- \quad \text{for} \quad x < 0; \; v(x) = u_+ \quad \text{for} \quad x > 0 \tag{1.11}$$

it has the discontinuous solution

$$u(x,t) = u_- \text{ for } x < st; \; u(x,t) = u_+ \text{ for } x > st \tag{1.12}$$

corresponding to a shock traveling at the velocity s given by (1.8). In the case where f' is increasing on the segment going from u_- to u_+ it has also a continuous solution defined by

$$u(x,t) = u_- \quad \text{for} \quad x < f'(u_-)t \tag*{$(1.13)_1$}$$

$$f'(u(x,t)) = \frac{x}{t} \quad \text{for} \quad f'(u_-)t < x < f'(u_+)t \tag*{$(1.13)_2$}$$

$$u(x,t) = u_+ \quad \text{for} \quad f'(u_+)t < x . \tag*{$(1.13)_3$}$$

In that case we want to rule out the discontinuous solution (1.12) as nonphysical.

2. DISCONTINUITIES; PART TWO. ([2], 1957)

Mathematicians first focused their attention on the case of a scalar hyperbolic equation. The first attempts were related to a model equation introduced by Burgers ([12], 1948)

$$u_t + uu_x = \mu u_{xx} \ . \tag{2.1}$$

Hopf ([13], 1950) and Cole ([14], 1951) studied the limit case $\mu \to 0$ by transforming (2.1) into the heat equation but Oleinik succeeded first in proving general results for the viscosity method ([15], 1954; [16], 1955; [17], 1956).

The importance of entropy conditions to ensure uniqueness was recognized. Germain and Bader ([18], 1953) solved the case of Burgers' equation and Oleinik ([19], 1956) treated the general scalar case (1.7).

Lax ([20], 1952) started a pioneering work in studying difference schemes and developing the concepts for systems ([21], 1954); by his study of the Riemann problem and introduction of a shock admissibility criterion he opened the way towards understanding the case of systems (1.9) ([2], 1957).

Lax's solution of the Riemann problem in a general setting can be summarized as follows:

A system (1.9) is said to be strictly hyperbolic if the matrix $F'(U)$ has real distinct eigenvalues $\lambda_j(U)$ (increasing with j) with right and left eigenvectors $r_j(U)$ and $l_j(U)$.

The solution of the Riemann problem requires the description of shock curves: if U_- is given, the set of U_+ satisfying (1.10) for some s is locally made of p curves.

For each j one of these curves has tangent $r_j(U_-)$ at U_- and Lax's shock admissibility criterion for such a j-shock is to keep only the part satisfying

$$\lambda_j(U_-) > s > \lambda_{j-1}(U_-) \quad \text{and}$$
$$\lambda_{j+1}(U_+) > s > \lambda_j(U_+) \ . \tag{2.2}$$

The solution of the Riemann problem requires also the

solution of the ordinary differential equation

$$\frac{dV(\tau)}{d\tau} = r_j(V(\tau)) \tag{2.3}$$

which defines another curve with tangent $r_j(U_-)$ at U_- (it has a contact of class C^2 with the corresponding j-shock curve). A function w is called a j-Riemann invariant if it satisfies $w'(V) \cdot r_j(V) \equiv 0$; it is then constant on each curve (2.3).

The analog of (1.13) corresponds to the possibility of going from U_- to U_+ along one of these curves with $\lambda_j(U)$ increasing. The monotonicity of λ_j depends upon the sign of the following quantity

$$\kappa_j = (l_j(U), F''(U).(r_j(U), r_j(U))\ . \tag{2.4}$$

Genuine nonlinearity of the j-field is the assumption that κ_j is never 0 while linear degeneracy corresponds to $\kappa_j \equiv 0$ in which case (2.3) redefines the curves of j-shocks (contact discontinuities).

Under the assumption that each field is either genuinely nonlinear or linearly degenerate, the solution of the Riemann problem can be found, if U_+ is near U_-, by following pieces of the preceding curves in the right order.

3. DISCONTINUITIES; PART THREE. ([3], 1983)

In order to prove convergence of approximations, one needed estimates. Such estimates involving total variation were established first by Oleinik ([22], 1957) for a scalar equation: for a smooth initial data v the solution of

$$u_t + f'(u)u_x - \varepsilon u_{xx} = 0 \tag{3.1}$$

satisfies

$$\|u_x(.,t)\|_{L^1} \leq \|v_x\|_{L^1} \quad \text{for} \quad t > 0\ . \tag{3.2}$$

This estimate is kept for $\varepsilon = 0$ if one replaces the norms in (3.2) by the total variation of $u(.,t)$ and v.

Similar estimates for systems are not true and a careful analysis of the solution of the Riemann problem shows that the total variation of the solution may increase as a result of the interaction of shocks of different families.

The breakthrough came with the work of Glimm ([23], 1965). He used a finite difference scheme with a random choice and a term measuring future interaction of shocks was added to the total variation in order to obtain a nonincreasing functional. This method was the only one giving estimates on the total variation of the solution (assuming the data had a small enough total variation); soon it was used to prove other global existence results as in Nishida ([24], 1968), Nishida and Smoller ([25], 1973), Liu ([26], 1977) and to describe the asymptotic behaviour of solutions as in Glimm and Lax ([27], 1970), DiPerna ([28], 1975), Liu ([29], [30], 1977); later it was made deterministic by Liu ([31], 1977).

The problem of understanding which discontinuities were admissible was still a crucial issue. Oleinik solved the case of a scalar equation ([32], 1959): her shock admissibility condition involves the respective positions of the graph of f and the cord joining $(u_-,f(u_-))$ and $(u_+,f(u_+))$ in order to accept the discontinuous solution (1.12)

if $u_- < u_+$ the cord is below the graph $(3.3)_1$

if $u_- > u_+$ the cord is above the graph . $(3.3)_2$

The case of systems gave rise to many different criterions. Another shock admissibility criterion was introduced by Liu ([33], 1974), but most of the other criterions which were proposed mentioned viscosity or entropy: as pointed out by Dafermos, the umbilical cord that joins the theory of systems of conservation laws with continuum physics is still vital for the proper development of the subject and it should not be severed.

The viscosity admissibility criterion was based on the study of traveling waves of an associated system where a diffusion term, analogous to adding viscosity effects in some conservation laws of continuum mechanics, had been added: Conley and Smoller ([34], 1970) started an efficient collaboration in that direction.

In the case where U_- and U_+ satisfy (1.10), one looks for a solution of the form $V(\frac{x - st}{\varepsilon})$ of a regularized equation

$$U_t + (F(U))_x - \varepsilon D U_{xx} = 0 \tag{3.4}$$

which leads to the differential equation

$$F(V(\tau)) - sV(\tau) - D\,\frac{dV(\tau)}{d\tau} = F(U_\pm) - sU_\pm \tag{3.5}$$

and the question of existence of an orbit connecting U_- to U_+.

The entropy admissibility criterion was introduced by Lax ([35], 1971) and it applies to systems that admit a convex "entropy", as physical examples do: Friedrichs and Lax ([36], 1971).

For equation (1.9) an entropy ϕ and its associated entropy flux ψ are functions of U satisfying

$$\phi'(U).F'(U) \equiv \psi'(U) \tag{3.6}$$

which implies

$$(\phi(U))_t + (\psi(U))_x = 0 \quad \text{for smooth solutions of (1.9)} . \tag{3.7}$$

For equation (3.4) with $D = I$ (artificial viscosity) one deduces

$$(\phi(U))_t + (\psi(U))_x - \varepsilon(\phi(U))_{xx} + \varepsilon\phi''(U).(U_x,U_x) = 0 . \tag{3.8}$$

Lax's entropy admissibility criterion stipulates that only solutions satisfying (in the sense of measures)

$$(\phi(U))_t + (\psi(U))_x \leq 0 \tag{3.9}$$

for each convex entropy ϕ are acceptable. The relation (3.9) would follow from (3.8) if one had enough estimates in order to apply a compactness argument.

The entropy rate admissibility criterion was introduced by Dafermos ([37], 1973).

Even after all these proposals to characterize the physical solution that one was seeking, uniqueness was established only in special situations: DiPerna ([38], 1979), Dafermos ([39], 1979).

4. OSCILLATIONS; PART ONE. ([40], 1979)

Even if one did not know exactly what should characterize the physical solutions of the basic systems of conservation laws of continuum physics, there were classical attempts to approach them: either use a numerical scheme or use a parabolic regularization. A convergence problem had to be solved for both methods when either the mesh size or the "viscosity" coefficient was going to zero.

As I had learned from J. L. Lions ([41], 1969), there were two important (unrelated) arguments to perform this limiting procedure, a compactness or a monotonicity argument. Having enough estimates was a requirement for application of the compactness argument and only Glimm had found a way to obtain them for systems.

The monotonicity method did not require as many estimates but it applied mostly to elliptic equations. For some systems arising from continuum mechanics, e.g. nonlinear elasticity, the monotonicity method was even known to be useless: in the case of hyperelasticity, monotonicity assumptions would imply convexity of the stored energy functional, which is known to be incompatible with frame indifference.

I had never heard anything on nonlinear elasticity before attending a meeting in Marseille in September 1975 where J. Ball explained to me his results ([42], 1977). It was only then that I realized that the monotonicity argument could be replaced by the use of the div-curl lemma, which I had obtained with F. Murat before my stay in Madison (I had understood before how to use our lemma in simplifying arguments in homogenization so I showed it to J. L. Lions at that meeting: [43], 1976).

In the div-curl lemma one has an open set Ω of R^N and $2N$ sequences, denoted $u_{j\varepsilon}$ and $v_{j\varepsilon}$, converging weakly in $L^2(\Omega)$ to u_{j0} and v_{j0}. One then deduces that $\sum_j u_{j\varepsilon}v_{j\varepsilon}$ converges to $\sum_j u_{j0}v_{j0}$ in the sense of distributions, provided that one has enough information on $div(u_\varepsilon)$ and $curl(v_\varepsilon)$, namely that each one of the quantities $\sum_j \frac{\partial u_{j\varepsilon}}{\partial x_j}$, and $\frac{\partial v_{j\varepsilon}}{\partial x_k} - \frac{\partial v_{k\varepsilon}}{\partial x_j}$ for $j \neq k$, stay in a compact set of the space $H^{-1}_{loc}(\Omega)$.

I then tried to use the div-curl lemma to prove existence theorems for systems of nonlinear partial differential equations; in addition to nonlinear elasticity, I also was thinking of applications to hyperbolic systems of two equations. It took me some time to find and put together the different pieces of the puzzle. In the Spring of 1977 I had found how to use the method for Burgers' equation by proving that a sequence of smooth solutions converging in the weak * topology of L^∞ was indeed converging in a stronger sense, but, for discontinuous solutions, a technical problem remained: there was a term describing the variation of entropy that did not satisfy (in an obvious way) the conditions needed for applying the div-curl lemma. In the Fall of 1977 I visited the Mathematics Research Center (at the time of a similar Symposium) and there I found the missing link.

The critical term was bounded in the space of measures and this did not ensure that it stayed in a compact set of H^{-1}_{loc}; but this followed from the fact that this term was also bounded in another negative Sobolev space $W^{-1,\infty}$, a straightforward consequence of a lemma proven recently by F. Murat in a completely different problem.

Meanwhile I had explained the basic ideas in a series of lectures held in Paris in February-March 1977 (the notes of this "cours Peccot" have not been published) where I described both homogenization and compensated compactness. The name had been coined by Lions to describe an improvement that Murat had made from our div-curl lemma ([44], 1978) which I immediately generalized to give it the form it still has ([45], 1978).

I described soon after ([46], 1978) this special application and gave lectures at Edinburgh on the general principles of the method, the corresponding notes ([40],

1979) being written by B. Dacorogna (my careless proofreading left a big misprint on the description of what should be done for hyperbolic systems: on the last line of page 209, one should of course read $(\nu,\phi_1)(\nu,\psi_2) - (\nu,\phi_2)(\nu,\psi_1)$ instead of 0; it had been written correctly before in [46]). It was only much later ([47], 1981) that Murat decided to publish a proof of his lemma.

5. OSCILLATIONS; PART TWO. ([48], 1983)

In [40], oscillations were described in the spirit of the work of Young on generalized curves ([49], 1937; [50], 1969), based on the following properties.

If a sequence of functions u_n, defined on an open set Ω of R^N and taking values in a compact set K of R^p, is only converging weakly, then the possible weak limits of all the associated sequences $F(u_n)$ can be described with the help of a probability measure ν_x on K for almost every point x of Ω: for a subsequence u_m the weak limit f of $F(u_m)$ is then given by $f(x) = (\nu_x, F)$ almost everywhere, this being valid for every continuous function F. In this setting the compensated compactness method is based on the remark that some information on derivatives of u_n may imply additional constraints on the Young's measures ν_x: typically, if one denotes by $\underline{u}_x$ the center of mass of ν_x, one obtains inequalities $(\nu_x, F) \geq F(\underline{u}_x)$ for a particular class of nonconvex functions F defined on R^p.

The class of such functions F depends upon how much is known concerning the derivatives of u_n, but it is difficult to describe it in a general situation. Fortunately, the quadratic functions of this class have been studied and most of the applications of the method are based on their characterization (the subclass of functions F for which $(\nu_x, F) = F(\underline{u}_x)$ holds is often the only one considered).

At the beginning, the method was only a substitute to proving estimates. Without estimates on the approximating sequence u_n, I was pessimistic and I thought that oscillations could occur. After finding that there were constraints imposed on the Young's measures, I became optimistic hoping that they were so restrictive that only Dirac measures would satisfy them, what would imply the existence of a strongly convergence subsequence u_m.

By applying the div-curl lemma to two equations (3.8) corresponding to entropies ϕ_1 and ϕ_2 and using the Young measures (parametrized by x and t) I had deduced that each of these

$$(\nu, \phi_1\psi_2 - \phi_2\psi_1) = (\nu, \phi_1)(\nu, \psi_2) - (\nu, \phi_2)(\nu, \psi_1) \ . \qquad (5.1)$$

For a scalar equation every function ϕ is an entropy and, using the family $\phi_k(u) = (u - k)_+$, I found that each ν should have its support in an interval where f was affine.

After this successful application of the method to the case of a scalar hyperbolic equation it was natural to try to solve the case of systems in a similar way. The functional equation (5.1) required the use of many entropy functions and, even for $p = 2$ where an infinite family of such entropies exist as solutions of a second order linear hyperbolic equation, it was difficult to handle. In my opinion the method had to be improved.

DiPerna ([51], [52], 1983) was more confident, and having a better understanding of the subject of hyperbolic systems of conservation laws than me, he found the way to use (5.1) for some systems of two equations. He used for that purpose families of entropies given by an expansion

$$\phi_k = e^{kw}(V_0 + \frac{V_1}{k} + ..) \qquad (5.2)$$

for a Riemann invariant w.

I lectured again on the theory after DiPerna's result ([48], 1983), but I had been already working for some time in a slightly different direction.

6. OSCILLATIONS; PART THREE. ([53], 1984)

In 1977 my work was based on the idea that partial differential equations of continuum physics describing macroscopic quantities should be stable under weak convergence ([54], 1978).

If this was not so, some new quantities would have to be introduced (like the internal energy of a gas), expressing the mathematical remark that if a sequence of functions f_n converges weakly to f and its square f_n^2 converges to g then one has $g \geq f^2$.

I tested this idea on semilinear hyperbolic systems arising as discrete velocity models in the kinetic theory of gas but, to my surprise, these models were not stable. I characterized the subclass of semilinear hyperbolic systems which are stable: the nonaffine ones occur only in one space variable and have the following form

$$u_t^j + c^j u_x^j = Q^j(u), \quad j = 1..p \tag{6.1}$$

where Q^j are quadratic polynomials satisfying

$$\frac{\partial^2 Q^j}{\partial u_k \partial u_l} = 0 \quad \text{for all } k,l \text{ such that } c^k = c^l \, . \tag{6.2}$$

During my visit to the Mathematics Research Center in the Summer 1980 ([55], 1980) I found other interesting properties of these systems, which improved an earlier result of global existence for small data in $L^1(R)^p$, concerning asymptotic behaviour and existence of a scattering operator.

I studied also two classical examples which were not stable with respect to weak convergence: the Carleman model

$$u_t + u_x + u^2 - v^2 = 0 \tag{6.3$_1$}$$

$$v_t - v_x - u^2 + v^2 = 0 \tag{6.3$_2$}$$

and the Broadwell model

$$u_t + u_x + \frac{1}{\delta}(uv - w^2) = 0 \tag{6.4$_1$}$$

$$v_t - v_x + \frac{1}{\delta}(uv - w^2) = 0 \tag{6.4$_2$}$$

$$w_t - \frac{1}{\delta}(uv - w^2) = 0 \, . \tag{6.4$_3$}$$

I studied the properties of weak convergence of the solutions corresponding to oscillating sequences of initial data and I found that the problem was propagation and interaction of oscillations. The repeated use of the div-curl lemma gave me a quite complete understanding for the Carleman model ([56], 1981): knowing the Young measures describing the oscillations for the initial data one could deduce the corresponding Young measures describing the oscillations of the solutions. The same method gave only an incomplete description of interaction of oscillations for the Broadwell model: some correlations of oscillations were necessary.

A year later, working with G. Papanicolaou, we solved completely the special case of oscillating data modulated in a periodic manner

$$\begin{aligned} &u(x,0) = U_0(x, \tfrac{x}{\varepsilon}), \quad v(x,0) = V_0(x, \tfrac{x}{\varepsilon}), \\ &w(x,0) = W_0(x, \tfrac{x}{\varepsilon}) \end{aligned} \tag{6.5}$$

where U_0, V_0 and W_0 have period 1 in their last argument. The oscillations are then described by functions U,V,W (depending on x,y,t) satisfying the following system

$$U_t + U_x + \frac{1}{\delta}\left(U \int_0^1 V(x,z,t)dz - \int_0^1 W^2(x,z,t)dz\right) = 0 , \qquad (6.6)_1$$

$$V_t - V_x + \frac{1}{\delta}\left(V \int_0^1 U(x,z,t)dz - \int_0^1 W^2(x,z,t)dz\right) = 0 , \qquad (6.6)_2$$

$$W_t - \frac{1}{\delta}\left(\int_0^1 U(x,y - z,t)V(x,y + z,t)dz - W^2\right) = 0 . \qquad (6.6)_3$$

As the general problem is not yet entirely understood, we only described these preliminary results much later ([53], 1984; [57], 1985).

7. ARE OSCILLATIONS REAL?

We have seen already oscillations in hyperbolic systems in two different situations. The first one was when oscillations present in the initial data were wiped out almost instantaneously by a nonlinear effect.

This was shown to occur, in the scalar case [2], [40] and then for some systems of two equations [51], [52], under a hypothesis of genuine nonlinearity. The second one was when oscillations were able to propagate and interact. This was the case for semilinear systems [53], [56], [57].

A third situation could exist where, even without any oscillations present in the initial data, some may develop later.

I was thinking that an example of this situation could appear in studying the kinetic theory of gas. When the mean free path between two collisions goes to zero one usually writes formally some fluid approximation (Hilbert or Chapman and Enskog expansions). In the case of the Broadwell system (6.5), it consists in studying the limit case $\delta \to 0$. If one defines mass density ρ and momentum density q by

$$\rho = u + v + 2w \tag{7.1$_1$}$$

$$q + u - v \tag{7.1$_2$}$$

one has the two conservation equations

$$\rho_t = q_x = 0 \tag{7.2$_1$}$$

$$q_t + (u + v)_x = 0 \,. \tag{7.2$_2$}$$

If $uv = w^2$ then $u + v$ is a quantity $F(\rho,q)$ and the usual formal guess is that, in the limit $\delta \to 0$, one should replace $(7.2)_2$ by

$$q_t + F(\rho,q)_x = 0 \,. \tag{7.3}$$

It is not known if this procedure is mathematically correct in general (Caflisch [58], 1983). From the H-theorem and the inequality $(b - a)(\mathrm{Log}b - \mathrm{Log}a) > 2(\sqrt{b} - \sqrt{a})^2$, one deduces that w and $\sqrt{uv}$ have the same oscillations when $\delta \to 0$. The formal procedure will be wrong if oscillations were occurring and we lack estimates to rule them out.

From this point of view, I suggest that oscillations are necessary, exactly in the same manner that Stokes proposed discontinuities as a necessity for obtaining solutions defined for all time.

When I try to learn more about Physics through reading what physicists teach, I am amazed to see the strange rules that they have invented in order to circumvent the huge difficulties encountered for explaining the unusual properties of elementary particles. I believe now that the reason of this curious behaviour is related to the presence of oscillating solutions and, as these objects are not yet well understood, it was necessary for them to guess some rules. Most of my ideas about this interpretation are related to homogenization (also in joint work with F. Murat) and it would be too long to describe them here; anyway, I have not yet succeeded in creating an improved mathematical tool for studying general oscillations in nonlinear partial differential equations [59].

When I started, in the Summer 1980, to study oscillations for the Carleman and the Broadwell models, I did not know what I was going to find and I could not have discussed then about three different situations as I do now.

Strangely enough an example of the third situation had already been published a year before by Lax and Levermore ([60], 1979). Their analysis relied on the inverse scattering method for the Korteweg de Vries equation. It is then a very special case and their detailed computations ([61], [62], [63], 1983) have not been generalized yet to any other equation.

A long time after hearing a qualitative description of their result and reading their first article I realized that they had found what I had been looking for even before I had started.

8. CONCLUSION.

So oscillations do exist.

Although some of my friends have been very optimistic about the possible use of my ideas, I believe that a better mathematical tool is necessary to study general oscillations in nonlinear partial differential equations.

My attempts to create this tool are based on different ideas which appeared in work on homogenization and I plan to describe them soon.

REFERENCES

1. Courant, R. and Friedrichs, K. O., Supersonic Flow and Shock Waves. Interscience, New York (1948), Applied Mathematical Sciences 21, Springer (1976).
2. Lax, P. D., "Hyperbolic systems of conservation laws II," Comm. Pure Appl. Math. 10 (1957), 537-566.
3. Dafermos, C. M., "Hyperbolic systems of conservation laws," in Systems of Nonlinear Partial Differential Equations, Ball, J. M., ed., NATO ASI Series C 111, Reidel, Dordrecht (1983), 25-70.
4. Challis, J., "On the velocity of sound," Philos. Magasine 3 32 (1848), 494-499.
5. Poisson, S. D., "Mémoire sur la théorie du son," J. Ecole Polytechnique, 14ème cahier 7 (1808), 319-392.
6. Stokes, E. E., "On a difficulty in the theory of sound," Philos. Magasine 3 33 (1848), 349-356.
7. Earnshaw, S., "On the mathematical theory of sound," Trans. Roy. Soc. of London 150 (1860), 133-148.
8. Riemann, B., "Über die Fortpssanzung ebener Luftwellen von endlicher Schwingungsweite," Abh. Gesellsch. Wiss. Gôttingen Math.-Phys. Kl. 8 43 (1860). Gesammelte Werke (1876), p. 144.
9. Rankine, W. J. M., "On the thermodynamic theory of waves of finite longitudinal disturbance," Trans. Roy. Soc. London 160 (1870), 277-288.
10. Hugoniot, H., "Sur la propagation du mouvement dans les corps et spécialement dans les gaz parfaits," J. Ecole Polytechnique 58 (1889), 1-125.
11. Rayleigh, J., "Aerial waves of finite amplitude," Proc. Roy. Soc. 84 (1910), 247-284. Scientific Papers vol. V, 573-610, Cambridge University Press (1912).
12. Burgers, J. M., "A mathematical model illustrating the theory of turbulence," Advances in Appl. Mech. 1 (1948), 171-199.
13. Hopf, E., "The partial differential equation $u_t + uu_x = \mu u_{xx}$," Comm. Pure Appl. Math. 3 (1950), 201-230.

14. Cole, J. D., "On a quasilinear parabolic equation occurring in aerodynamics," Quart. Appl. Math. 9 (1951), 226-236.

15. Lax, P. D., "The initial value problem for nonlinear hyperbolic equations in two independent variables," Ann. Math. Studies Princeton 33 (1954), 211-229.

16. Oleinik, O. A., "On Cauchy's problem for nonlinear equations in the class of discontinuous functions," Uspehi Mat. Nauk 9 (1954), 231-233.

17. Oleinik, O. A., "Boundary problems for partial differential equations with small parameters in the highest order and Cauchy's problem for nonlinear equations," Uspehi Mat. Nauk 10 (1955), 229-234.

18. Oleinik, O. A., "On discontinuous solutions of nonlinear differential equations," Dokl. Akad. Nauk SSSR 109 (1956), 1098-1101.

19. Germain, P. and Bader, R., "Unicité des écoulements avec chocs dans la mécanique de Burgers," O.N.E.R.A. Paris (1953).

20. Lax, P. D., "On discontinuous initial value problems for nonlinear equations and finite difference schemes," L.A.M.S. 1332 Los Alamos (1952).

21. Lax, P. D., "Weak solutions of nonlinear hyperbolic equations and their numerical computation," Comm. Pure Appl. Math. 7 (1954), 159-193.

22. Oleinik, O. A., "Discontinuous solutions of nonlinear differential equations," Uspehi Mat. Nauk 12 3 (1957), 3-73. AMS Transl. Ser. 2 26, 95-172.

23. Glimm, J., "Solutions in the large for nonlinear hyperbolic systems of equations," Comm. Pure Appl. Math. 18 (1965), 697-715.

24. Nishida, T., "Global solution for an initial boundary value problem of a quasilinear hyperbolic system," Proc. Japan Acad. 44 (1968), 642-646.

25. Nishida, T. and Smoller, J. A., "Solutions in the large for some nonlinear hyperbolic conservation laws," Comm. Pure Appl. Math. 26 (1973), 183-200.

26. Liu, T. P., "Initial-boundary value problems for gas dynamics," Arch. Rational Mech. Anal. 64 (1977), 137-168.

27. Glimm, J. and Lax, P. D., "Decay of solutions of systems of nonlinear hyperbolic conservation laws," Mem. Amer. Math. Soc. 101 (1970).

28. DiPerna, R. J., "Decay and asymptotic behaviour of solutions to nonlinear hyperbolic systems of conservation laws," Indiana Univ. Math. J. 24 (1975), 1047-1071.

29. Liu, T. P., "Decay to N-waves of solutions of general systems of nonlinear hyperbolic conservation laws," Comm. Pure Appl. Math. 30 (1977), 585-610.

30. Liu, T. P., "Linear and nonlinear large-time behaviour of solutions of hyperbolic conservation laws," Comm. Pure Appl. Math. 30 (1977), 767-796.

31. Liu, T. P., "The deterministic version of the Glimm scheme," Comm. Math. Phys. 57 (1977), 135-148.

32. Oleinik, O. A., "Uniqueness and stability of the generalized solution of the Cauchy problem for a quasi-linear equation," Uspehi Mat. Nauk 14 (1959), 165-170.

33. Liu, T. P., "The Riemann problem for general 2×2 conservation laws," Trans. Amer. Math. Soc. 199 (1974), 89-112.

34. Conley, C. C. and Smoller, J. A., "Viscosity matrices for two dimensional nonlinear hyperbolic systems," Comm. Pure Appl. Math. 23 (1970), 867-884.

35. Lax, P. D., "Shock waves and entropy," in Contributions to Nonlinear Functional Analysis, Zarantonello, E. H., ed., Academic Press, New York (1971), 603-634.

36. Friedrichs, K. O. and Lax, P. D., "Systems of conservation equations with a convex extension," Proc. Nat. Acad. Sci. U.S.A. 68 (1971), 1686-1688.

37. Dafermos, C. M., "The entropy rate admissibility criterion for solutions of hyperbolic conservation laws," J. Differential Equations 14 (1973), 202-212.

38. DiPerna, R. J., "Uniqueness of solutions of hyperbolic conservation laws," Indiana Univ. Math. J. 28 (1979), 137-188.

39. Dafermos, C. M., "The second law of thermodynamics and stability," Arch. Rational Mech. Anal. 70 (1979), 167-179.

40. Tartar, L. C., "Compensated compactness and applications to partial differential equations," in Nonlinear Analysis and Mechanics; Heriot-Watt Symposium vol. IV, Knops, R. J., ed., Research Notes in Mathematics 39, Pitman, London (1979), 136-212.

41. Lions, J. L., Quelques Méthodes de Résolution des Problèmes aux Limites Non Linéaires, Dunod, Paris (1969).

42. Ball, J. M., "Convexity conditions and existence theorems in nonlinear elasticity," Arch. Rational Mech. Anal. 63 (1977), 337-403.

43. Lions, J. L., "Asymptotic behaviour of solutions of variational inequalities with highly oscillating coefficients," in Applications in Functional Analysis to Problems in Mechanics, Germain, P. and Nayrolles, B., eds., Lecture Notes in Mathematics 503, Springer, Berlin (1976), 30-55.

44. Murat, F., "Compacité par compensation," Ann. Scuola Norm. Sup. Pisa 5 (1978), 489-507.

45. Tartar, L. C., "Une nouvelle méthode de résolution d'équations aux dérivées partielles nonlinéaires," in Journées d'Analyse Non Linéaire, Bénilan, P. and Robert, J., eds., Lecture Notes in Mathematics 665, Springer, Berlin (1978), 228-241.

46. Tartar, L. C., "Equations hyperboliques non linéaires," Séminaire Goulaouic Schwartz (1977-1978), exp. XVIII, Ecole Polytechnique, Palaiseau (1978).

47. Murat, F., "L'injection du cône positif de H^{-1} dans $W^{-1,q}$ est compacte pour tout q 2," J. Math. Pures Appl. 60 (1981), 309-322.

48. Tartar, L. C., "The compensated compactness method applied to systems of conservation laws," in Systems of Nonlinear Partial Differential Equations, Ball, J. M., ed., NATO ASI Series C 111, Reidel, Dordrecht (1983), 263-285.

49. Young, L. C., "Generalized curves and the existence of an attained absolute minimum in the calculus of variations," C. R. Soc. Sci. Lett. Varsovie, Classe III 30 (1937), 212-234.

50. Young, L. C., Lectures on the Calculus of Variation and Optimal Control Theory, W. B. Saunders, Philadelphia (1969).

51. DiPerna, R. J., "Convergence of approximate solutions to conservation laws," Arch. Rational Mech. Anal. 82 (1983), 27-70.

52. DiPerna, R. J., "Convergence of the viscosity method for isentropic gas dynamics," Comm. Math. Phys. 91 (1983), 1-30.

53. Tartar, L. C., "Etude des oscillations dans les équations aux dérivées partielles non linéaires," in Trends and Applications of Pure Mathematics to Mechanics, Roseau, M. and Ciarlet, P. G., eds., Lecture Notes in Physics 195, Springer, Berlin (1984), 384-412.

54. Tartar, L. C., "Nonlinear constitutive relations and homogenization," in Contemporary Developments in Continuum Mechanics and Partial Differential Equations, de La Penha, G. M. and Medeiros, L. A., eds., Mathematics Studies 30, North Holland, Amsterdam (1978), 472-484.

55. Tartar, L. C., "Some existence theorems for semilinear hyperbolic systems in one space variable," Mathematics Research Center Technical Summary Report #2164, University of Wisconsin-Madison, Madison, WI (1980).

56. Tartar, L. C., "Solutions oscillantes des équations de Carleman," Goulaouic Meyer Schwartz Seminar (1980-1981), exp. XII, Ecole Polytechnique, Palaiseau (1981).

57. McLaughlin, D., Papanicolaou, G. and Tartar, L. C., "Weak limits of semilinear hyperbolic systems with oscillating data," in Macroscopic Modelling of Turbulent Flows, Frisch, U., Keller, J. B., Papanicolaou, G. and Pironneau, O., eds., Lecture Notes in Physics 230, Springer, Berlin (1985), 277-289.

58. Caflisch, R. E., "Fluid dynamics and the Boltzmann equation," in Nonequilibrium Phenomena. I. The Boltzmann Equation, Lebowitz, J. L. and Montroll, E. W., eds., Studies in Statistical Mechanics vol. X, North-Holland, Amsterdam (1983), 193-223.

59. Tartar, L. C., "Oscillations in nonlinear partial differential equations: compensated compactness and homogenization," to appear in Lecture Notes in Applied Mathematics 23, American Mathematical Society, Providence (1986).

60. Lax, P. D. and Levermore, C. D., "The zero dispersion limit for the Korteweg de Vries equation," Proc. Nat. Acad. Sci. U.S.A. 76 8 (1979), 3602-3606.

61. Lax, P. D. and Levermore, C. D., "The small dispersion limit of the Korteweg de Vries equation I," Comm. Pure Appl. Math. 36 (1983), 253-290.

62. Lax, P. D. and Levermore, C. D., "The small dispersion limit of the Korteweg de Vries equation II," Comm. Pure Appl. Math. 36 (1983), 571-593.

63. Lax, P. D. and Levermore, C. D., "The small dispersion limit of the Korteweg de Vries equation III," Comm. Pure Appl. Math. 36 (1983), 809-830.

Centre d'Etudes de Limeil-Valenton
DMA/MCN
BP. 27
94190 VILLENEUVE SAINT GEORGES
FRANCE

The Structure of Manifolds with Positive Scalar Curvature

Richard Schoen and Shing-Tung Yau

In our study of the positive mass conjecture [5] and positive action conjecture [6] in general relativity, we came across the problem of determining the topology of complete manifolds with positive scalar curvature. For complete three-dimensional manifolds with positive scalar curvature, the results were published in [7] and [9]. For complete manifolds with dimension greater than three, the first result appeared in [8]. Later, some of these results were reproved by Gromov-Lawson in [1] using arguments of harmonic spinors which generalize the works of Lichnerowicz [4] and Hitchin [3]. Soon afterwards, we were able to give a more complete understanding of complete manifolds with positive scalar curvature. Both of us have lectured on these works in the past few years. We summarize some of these results below.

First of all, we remark that the argument in [9] shows the following:

DIRECTIONS IN PARTIAL
DIFFERENTIAL EQUATIONS

ISBN 0-12-195255-X

Theorem 1.

Let M be a complete orientable three-dimensional manifold with scalar curvature ≥ 1. Then we can write M as an increasing union of compact subdomains Ω_1, each of which is diffeomorphic to the complement of a finite number of disjoint balls of a three-dimensional manifold of the form $M_1\#\ldots\#M_k\#N_1\#\ldots\#N$, where "#" means connected sum, $M_1 = S^2 \times S^1$ and N_i is a compact three-dimensional manifold with finite fundamental group.

We remark that it is a conjecture in topology that a compact three-dimensional manifold with finite fundamental group is a spherical space form. If this conjecture is true, Theorem 1 gives a complete classification of complete three-dimensional manifold with scalar curvature ≥ 1. The last assumption can be replaced by requiring only scalar curvature ≥ 0. Under this weaker assumption, we have to allow handlebodies as possible connected summands in the above theorem. In this way, we can even allow our three-dimensional manifold to have compact boundaries whose mean curvature is nonnegative with respect to the outward normal.

As a consequence of the theorem in [9], we can also prove the following proposition which will be used in understanding manifolds with positive scalar curvature in higher dimension.

Proposition 1.

Let M be a complete three-dimensional manifold with scalar curvature ≥ 1. Then we can write M as an increasing union of compact subdomains Ω_1 so that each component of $\partial\Omega_1$ has an area less than some constant C which is independent of i.

As a theorem of independent interest, we mention the following:

Theorem 2.

Let M be a complete three-dimensional manifold which is connected at infinity. If the scalar curvature of M is greater than one, then there exists no distance-increasing map from the real line into M.

All of the above theorems have a finite version, i.e., we do not have to assume that M is complete and the conclusion is focused on the part of M which is not close to ∂M.

When dim M > 3, we proved the following theorem quite a while ago.

Theorem 3.

Let M be a compact manifold with positive scalar curvature. Then M cannot be represented as a homology class in a compact manifold with nonpositive sectional curvature.

We proved this theorem by following our arguments in [8] using minimal hypersurfaces. Note that the technical assumption of dimensions ≤ 7 in [8] was dropped by us a few years ago. Independently, Gao, in his Stony Brook thesis, was able to prove Theorem 3 following an argument of Gromov-Lawson. Some special cases of the following two theorems were also obtained by Gromov-Lawson independently.

Theorem 4.

Let M be an n-dimensional complete manifold with positive scalar curvature. Then there exists no distance-decreasing map from M onto R^{n-1} with the following property: For each ball B(r) of radius r in R^{n-1}, there exists a codimensional one hypersurface H in M which has a nonzero intersection number with the curve $F^{-1}(\emptyset)$ and $F(\partial H) \subset \partial B(r)$.

Theorem 5.

Let M be an n-dimensional complete manifold with scalar curvature ≥ 1. Then there exists no codimension two submanifold N in M with the following property. There exists a distance-decreasing map from N onto R^{n-2} and there exists no immersed two-dimensional sphere which has nonzero intersection number with N.

While Theorem 3, Theorem 4, and Theorem 5 were proved by us a long time ago, we feel the following theorem gives a more complete picture of manifolds with positive scalar curvature.

Theorem 6.

Let M be a compact four-dimensinal manifold with positive scalar curvature. Then there exists no continuous map with nonzero degree onto a compact $K(\pi,1)$.

The second author gave a detailed proof of Theorem 5 in the Summer Conference on Nonlinear Functional Analysis in Berkeley. It depends on the following statement which was published in Schoen-Yau [8]. Let Σ be a stable minimal surface in a three-dimensinal manifold with scalar curvature greater than one, then any point on Σ has a distance less than $\frac{5\pi}{3}$ from $\partial\Sigma$.

Lemma. Let $\tilde{M}$ be a complete manifold which covers a compact manifold M. Let N be a compact submanifold of M whose area is less than C. If N is homologous to zero in $\tilde{M}$, then N is the boundary of some compact orientable manifold P so that $\sup_{\in P} d(x,N)$ is less than some constant depending only on C and the geometry of M.

Combining Proposition 1 and the Lemma, we can outline the proof of Theorem 5 in the special case when M is a four-dimensional compact $K(\pi,1)$. We take a closed geodesic which represents a nontrivial element in $\pi_1(M)$. Lift it to the universal cover $\tilde{M}$ of M. Call this geodesic σ. Then we can find a properly embedded three-dimensional area minimizing hypersurface H in M which intersects σ at finite number of points with nonzero algebraic intersection number. As was shown in [8], we can conformally deform the metric of H so that the scalar curvature is greater than one. From the construction of H, we can assume that either H is compact or every tubular neighborhood of σ will intersect H in a compact set. If H is compact, it must represent a nonzero three-dimensional homology class of $\tilde{M}$ which is a contradiction. Otherwise, we can use Proposition 1 to write H as an increasing union of compact subdomains Ω_i so that each component of $\partial\Omega_i$ has area less than a constant C. Then the above Lemma shows that when i is large enough, we can cap off each component of $\partial\Omega_i$ by a compact orientable hypersurface which does not intersect σ. In this way, we form a compact orientable hypersurface which has nonzero intersection number with σ. In this way, we obtain another contradiction.

Remark. The above example shows that one may have a complete three-dimensional manifold with scalar curvature greater than one and yet the diameter of some component of the geodesic sphere of some fixed point is arbitrarily large. (This fact remains true even when one perturbs the geodesic sphere.) We believe that by refining the above argument, we can prove Theorem 5 for arbitrary dimension.

If a compact manifold with positive scalar curvature is also conformally flat, then its topology must be very restrictive. In fact, we believe that such manifolds must be diffeomorphic to the connected sum of compact manifolds which are covered by S^n, $S^{n-1} \times S^1$ or $S^{n-k} \times H^k$ where H^k is the hyperbolic space form with $n > 2k$. We have the following theorems which support this conjecture.

Theorem 7.

Let M be a n-dimensional complete manifold with nonnegative scalar curvature. Then any conformed immersion of M into S^n is one to one. In particular, any complete conformally flat manifold with nonnegative scalar curvatue is the quotient of a domain in S^n by a discrete subgroup of the conformal group.

Theorem 8.

Let M be a compact conformally flat manifold with positive scalar curvature. Then the universal cover of M is a domain Ω in S^n and the Hausedorff dimension of $S^n \setminus \Omega$ is less than $\frac{n}{2} - 1$. In particular, the homotopy groups $\pi_i(M) = \emptyset$ for $1 < i \leqslant \frac{n}{2}$.

Theorem 9.

Let M be a compact conformally flat manifold with non-negative scalar curvature so that M is the quotient of a domain by a discrete subgroup of the conformal group. Then unless M is covered by S^n, $S^{n-1} \times S^1$ on the torus, every element in the discrete group is conjugate to a dilation and the discrete group contains a free subgroup with more than two generators.

Theorem 10.

Let M be a compact manifold whose fundamental group is not of exponential growth. Then unless M is covered by S^n, $S^{n-1} \times S^1$ or the torus, M admits no conformally flat structure.

References

[1] M. Gromov and H.B. Lawson, Spin and scalar curvature in the presence of a fundamental group, Ann. of Math. 111 (1980), 209-230.

[2] M. Gromov and H.B. Lawson, Positive scalar curvature and the Dirac operator on complete Riemannian manifolds, Publ. IHES, No. 58 (1983).

[3] N. Hitchin, Harmonic spinors, Advances in Math. 14 (1974), 1-55.

[4] A. Lichnerowicz, Spineurs harmoniques, C.R. Acad. Sci. Paris, Ser. A-B, 257 (1963), 231-260.

[5] R. Schoen and S.T. Yau, Proof of the positive mass theorem II, Comm. Math. Phys. 79 (1981), 231-260.

[6] R. Schoen and S.T. Yau, Proof of the positive action conjecture in quantum relativity, Phys. Rev. Let. 42 (1979), 547-548.

[7] R. Schoen and S.T. Yau, Existence of incompressible minimal surfaces and the topology of manifolds with nonnegative scalar curvature, Ann. of Math. 110 (1979), 127-142.

[8] R. Schoen and S.T. Yau, On the structure of manifolds with positive scalar curvature, Manuscripta Math. 28 (1979), 159-183.

[9] R. Schoen and S.T. Yau, Complete three-dimensional manifolds with positive Ricci curvature and scalar curvature, Ann. of Math. 102 (1982), 209-228.

Department of Mathematics
University of California, San Diego
La Jolla, CA 92093

Index